An Economic Analysis of the Financial Records of al-Qa'ida in Iraq

Benjamin Bahney, Howard J. Shatz, Carroll Ganier,
Renny McPherson, Barbara Sude

With Sara Beth Elson, Ghassan Schbley

Prepared for the Office of the Secretary of Defense
Approved for public release; distribution unlimited

NATIONAL DEFENSE RESEARCH INSTITUTE

The research described in this report was prepared for the Office of the Secretary of Defense (OSD). The research was conducted in the RAND National Defense Research Institute, a federally funded research and development center sponsored by OSD, the Joint Staff, the Unified Combatant Commands, the Navy, the Marine Corps, the defense agencies, and the defense Intelligence Community under Contract W74V8H-06-C-0002.

Library of Congress Cataloging-in-Publication Data

An economic analysis of the financial records of al-Qa'ida in Iraq / Benjamin Bahney ... [et al.].
p. cm.
Includes bibliographical references.
ISBN 978-0-8330-5039-7 (pbk. : alk. paper)
1. Qaida (Organization)—Finance. 2. Terrorism—Iraq—Finance. 3. Terrorism—Finance. I. Bahney, Benjamin.

HV6433.I722Q357 2010
363.32509567—dc22

2010028999

Published 2010 by the RAND Corporation
1776 Main Street, P.O. Box 2138, Santa Monica, CA 90407-2138
1200 South Hayes Street, Arlington, VA 22202-5050
4570 Fifth Avenue, Suite 600, Pittsburgh, PA 15213-2665
RAND URL: http://www.rand.org/
To order RAND documents or to obtain additional information, contact
Distribution Services: Telephone: (310) 451-7002;
Fax: (310) 451-6915; Email: order@rand.org

Preface

This monograph conducts a series of economic analyses on two collections of documents detailing the administration of al-Qa'ida in Iraq (AQI) during 2005 and 2006 in Anbar province, Iraq. The documents contain information on the group's finances, payrolls, and organization.

The records show that AQI was a bureaucratic, hierarchical organization that exercised tight financial control over its largely criminally derived revenue streams. Using attack reports and AQI's spending data, we find that attacks statistically increased at a rate of one per every $2,700 sent by the provincial administration to a sector group within the province, suggesting that AQI paid not only for materiel but also for important expenses related to organizational sustainment, such as compensation and rents. AQI members were paid less and simultaneously faced a much greater risk of death than the general Anbar population, which suggests that financial rewards were not a primary reason for joining the group. Our findings imply that financial interdiction is one useful tool for slowing and disrupting militant attacks, but that neither interdiction nor material demobilization incentives on their own will be effective in ending a militant group.

This research was conducted within the Intelligence Policy Center of the RAND National Defense Research Institute, a federally funded research and development center sponsored by the Office of the Secretary of Defense, the Joint Staff, the Unified Combatant Commands, the Navy, the Marine Corps, the defense agencies, and the defense Intelligence Community.

For more information on the RAND Intelligence Policy Center, see http://www.rand.org/nsrd/about/intel.html, or contact the director (contact information is provided on the web page).

Contents

Figures

Tables

Summary

Terrorist and insurgent groups, both of which may be termed "militant groups," are economic actors: They have a fundamental need to mobilize resources. However, there has been relatively little research about the economic and financial decisionmaking of such groups based on actual financial records. With U.S., North Atlantic Treaty Organization, and other allied troops and governments engaged in counterinsurgency and counterterrorism operations, such research could help mitigate the threats posed by these groups by improving our understanding of their financial decisionmaking.

This monograph analyzes the finances of the militant group al-Qa'ida in Iraq (AQI) in Anbar province during 2005 and 2006, at the peak of the group's power and influence. We draw on captured financial records that recorded the daily financial transactions of both one specific sector within Anbar province and the AQI provincial administration. To our knowledge, this monograph offers one of the most comprehensive assessments of the financial operations of AQI or any other contemporary Islamic militant group.

Key Findings

AQI was a hierarchical organization with decentralized decisionmaking. The memos and financial ledgers in this collection of documents clearly indicate that AQI in Anbar province had a hierarchically organized system of financing and administration, with established bureaucratic relationships and rules. Although a hierarchy of administrative units collected reporting from the field, controlled the allocation of resources, and broadly administered the policies and procedures of AQI in Anbar, this arrangement is not inconsistent with the idea that a network of local commanders implemented these policies and made their own tactical decisions in a decentralized manner.

AQI in Anbar was profitable enough to send substantial revenues out of the province in 2006. News reports in 2006 and 2007 indicated that AQI was profitable enough to be financially self-sufficient, and that it was sending excess revenues to al-Qa'ida senior leaders in Pakistan. There was no explicit record of transfers to Pakistan in these documents, but there was evidence to suggest that the Anbar chief administrative officer, known as the administrative emir, exported revenues to other provinces in Iraq or to other countries.

AQI relied on extortion, theft, and black market sales to fund its operations in Anbar. AQI relied to a great extent on simple theft and resale, primarily of high-value items such as generators and cars, from Shi'a transiting Anbar province and people cooperating with Coalition Forces (CF). The group increased its reliance on the threat of violence to generate revenue in mid-2006, which we believe helped to turn influential tribes against it and toward collaboration with CF.

AQI needed large, regular revenue sources to fund its operations, but its administrative leaders did not hold much cash on hand. AQI required substantial funding to conduct its day-to-day operations, and the preponderance was needed to pay salaries. However, the financial records show that cash moved very rapidly; the administrative emirs carried only two weeks' worth of funds.

Data on compensation practices and risk of death indicate that AQI members were not compensated for their dramatically higher fatality rate. Individual members of AQI made less money than ordinary Anbaris—AQI average annual household compensation was $1,331 compared to $6,177 for average Anbar households—but faced a nearly 50-fold increase in the yearly risk of violent death. AQI compensation included monthly payments for members and their dependents, as well as monthly payments to the families of imprisoned and deceased members. These latter payments constituted a form of insurance unavailable to civilian Anbar households, but still resulted in lower risk-adjusted expected lifetime earnings. This is not to say that potential AQI recruits carefully computed their lifetime incomes. Rather, our results suggest that if AQI members were rational in their decisionmaking, financial rewards were not among the primary reasons for why they joined the group. Instead, other reasons must have been predominant to the point of being worth many years of forgone income and could have included ideological, religious, political, or nationalistic beliefs; tribal issues; matters of personal honor or revenge; or the simple desire for notoriety.

One-time payments are not sufficient to conduct either simple or complex attacks. Individual attacks were expensive, as AQI, like a firm, carried overhead costs, many of which were recurring. Although it may not have cost much to obtain the materiel used to carry out attacks, AQI incurred other recurring costs. When administrative costs, such as paying group members and the families of the imprisoned and deceased, securing and maintaining safe houses, and transporting materiel and members, were spread across the number of events, the cost of attacks was in the thousands of U.S. dollars. Our best estimate is that, on average, an additional attack cost the group $2,700. This amount is equivalent to 40 percent of the average household income.

Disrupting AQI's financial flows could disrupt the pace of attacks. Given the high cost of attacks, salaries, support to dependents, the rapidity with which money moved through the organization, the close correlation we find between funding and attacks, and the lack of substantial cash reserves, it is clear that AQI is highly sensitive to cash flows. Interrupting these flows may interrupt the number of attacks the organization can muster.

Implications

Captured financial ledgers may warrant greater emphasis as a source of strategic intelligence on militant groups. Given the important information gathered from just nine documents, including four financial records, militant financial records could be a particularly effective source of intelligence information that could yield strategic insights. In the case of AQI in Anbar, these documents revealed information about how the group was organized, how it made decisions, how it raised money, and what its financial vulnerabilities were. They also provided information with which we could analyze risks and rewards for members, with the result suggesting that financial motivations do not play a substantial role in motivating group membership.

The structure of a militant group's back office operations may reveal novel vulnerabilities. AQI was organized as a hierarchical group with decentralized decisionmaking. This system of organization makes it vulnerable. First, the vast paper trail it left behind gave CF an opportunity to track its activities. Second, eliminating the upper echelons of the administrative structure can at least temporarily disrupt operations and the administration of the group, although this effect may not prove readily apparent because of the autonomy of lower-level elements. In addition, some of these lower-level officials will have developed the skills to occupy positions higher up in the hierarchy in an effective manner. However, decentralization makes it easier to negotiate with lower levels of the organization, which could win over the group's periphery if lower-level members find inducements to halt fighting more attractive than what the center can offer, given the high level of risk involved in militancy.

A militant group's participation in the local political economy can be a key vulnerability. Combined with findings by other sources on the history of the Anbar insurgency, our analysis suggests that insurgent involvement in an economy can alienate other groups. In the case of AQI, the group competed with key tribes for revenues. Tribes eventually banded against AQI and allied themselves with the security forces in the interest of self-preservation. Collaboration between the tribes and the security forces led to the creation of new tribal militias and a larger, more capable force posture. These forces constrained AQI's ability to operate in the open, degrading the group's ability to sustain its funding. Although AQI's documents suggest that this process began to have a financial effect on the group starting in October 2006, only access to AQI records in Anbar for 2007 would enable further research to confirm or discredit this hypothesis in this specific instance.

The U.S. government's concept of "threat finance" intelligence should be broadened to "threat economics." Insurgency theorists, military operators, and intelligence officials have posited that the financing of insurgent groups is pivotal for sustaining their operations, and thus their financial systems should be key targets in operations by counterinsurgents. The wide acceptance of this principle has led to a focus on "threat finance," focusing on how threat groups raise money. Although the analysis provided in this monograph supports and potentially strengthens the notion that militant finances should be tracked, our findings also suggest that the notion of threat "finance" may be too narrow. Focusing on the revenue side alone ignores half of the useful data on militant back office operations. Our analysis indicates that understanding the spending patterns of militant groups may provide even more useful insights than understanding their revenue streams. We therefore recommend that the concept of "threat finance" be broadened to "threat economics" to better frame the attention of analysts, operators, commanders, and policymakers.

Acknowledgments

The authors are grateful to a number of people for their assistance in the course of this research. The research was only possible thanks to the vision of LtCol. Drew Cukor of the U.S. Marine Corps. Lt Col Steve Kiser of the U.S. Air Force was indispensible for his ideas and his support of our research. We also thank Jack Riley of RAND's National Security Research Division and John Parachini and Katharine Webb of RAND's Intelligence Policy Center for their exceptional support of this project and for helping to guide it to its conclusion. Extended conversations with Austin Long, Brian Gordon, Chris Meyer, MAJ Tim Blanch, Maj. Leo Gregory, Richard May, and Maj. Ray Gerber advanced our thinking on this topic and strengthened our research. We are deeply grateful to Keith Crane and Ben Connable for their thoughtful reviews. Jacob Shapiro helped to broaden our perspective on terrorist financing and was invaluable in directing us to useful data and sources. Staff members associated with the Iraq Body Count were also helpful in guiding us to useful data on noncombatant casualties in Anbar province. The authors also thank RAND colleagues James Bruce, Kim Cragin, Stephen Hosmer, Terrence Kelly, Alireza Nader, and William Rosenau for their assistance and insights. We are also grateful to Todd Duft, Carol Earnest, Stephen Kistler, Rena Rudavsky, and Maritta Tapanainen for their assistance with the publications process, and to Patricia Bedrosian for her careful and constructive editing.

Abbreviations

AQI	al-Qa'ida in Iraq
CF	Coalition Forces
COIN	counterinsurgency
COSIT	Central Organization for Statistics and Information Technology
CPA	Coalition Provisional Authority
CPI	consumer price index
CTC	Combating Terrorism Center
DF	direct fire
FATF	Financial Action Task Force
GDP	gross domestic product
GoI	government of Iraq
IA	Iraqi Army
IBC	Iraq Body Count
IED	improvised explosive device
I MEF	I Marine Expeditionary Force
ISF	Iraqi Security Forces

ISI	Islamic State of Iraq
JTJ	Jama'at al-Tawhid wa al-Jihad
KRSO	Kurdistan Region Statistics Office/Organization
MNF-I	Multi-National Force–Iraq
MNF-W	Multi-National Force–West
MSC	Mujahidin Shura Council
SAA	Sahwat al-Anbar ("the Anbar Awakening")
SIGACTS	Significant Activities (database)
SIGIR	Special Inspector General for Iraq Reconstruction
VBIED	vehicle-borne improvised explosive device

CHAPTER ONE

Introduction

Without exception, terrorist and insurgent groups are economic actors: They need to mobilize resources to conduct operations.[1] The close of the Cold War brought about a "new" version of terrorism. This contemporary version was driven by both the decline in state sponsorship of terrorist groups and the rapid emergence of globalization, leading groups toward increased independence from states and financial self-sufficiency.[2] Since then, both policymakers and academics have largely conflated terrorist and insurgent groups, both of which we term "militant groups." As a result, the study of how militant groups fund themselves now falls under the rubric of "terrorist financing." In the pages that follow, we will use this conventional term while acknowledging that the group we study fits some of the definitions of both a terrorist and an insurgent group.[3]

---

[1] James Adams, *The Financing of Terror,* New York: Simon & Schuster Press, 1986; R. T. Naylor, "The Insurgent Economy: Black Market Operations of Guerilla Organizations," *Crime, Law and Social Change,* Vol. 20, 1993, pp. 13–51.

[2] Jeanne K. Giraldo and Harold A. Trinkunas, "The Political Economy of Terrorism Financing," 2007b, in Jeanne K. Giraldo and Harold A. Trinkunas, eds., *Terrorism Financing and State Responses: A Comparative Perspective,* Stanford, Calif.: Stanford University Press, 2007a; Steven Simon and Daniel Benjamin, "America and the New Terrorism," *Survival,* Vol. 42, No. 1, Spring 2000, pp. 59–75.

[3] These definitions include large group size, use of subversion against the state, desire to control territory, and desire to defeat and replace the existing government. However, al-Qa'ida in Iraq not only had the intermediate goal to overthrow the government of Iraq, it also sought the broader goal of creating an Islamic caliphate.

This monograph analyzes the finances of one important militant group, al-Qa'ida in Iraq (AQI), in Anbar province during the peak of its power and influence, 2005 through 2006. We draw on captured financial documents that recorded both the daily financial transactions of one specific sector within Anbar province and the financial transactions by the overall provincial administration. Only a handful of studies use data reflecting the actual operations of militant groups. To our knowledge, this monograph offers one of the most comprehensive assessments of the financial operations of AQI or any other contemporary Islamic militant group.

Financing Militant Organizations

Although smaller groups typically use coercion, kidnapping, and extortion to fund themselves, large insurgencies rely on popular support, such as voluntary contributions, "revolutionary taxes," exploitation of natural resources, or extortion.[4] R. T. Naylor suggests that as a group's needs and capabilities increase, so does its tendency to adopt fund-raising techniques that are symbiotic with the formal economy. But at low and intermediate stages of militancy, militant groups conduct parasitical fund-raising activities that closely approximate those of organized crime.[5]

Theoretical studies have treated both organizational and also individual-level concerns regarding militant financing. Organizational studies suggest that there are two inherent principal-agent problems in financing militancy. First, militant groups face information problems because of the simultaneous need to be clandestine and to monitor the fund-raising tasks of specialized financing units. Recent work has identified this phenomenon in radical Islamic groups more specifically,

[4] Thomas X. Hammes, "Countering Evolved Insurgent Networks," *Military Review*, July–August 2006, pp. 18–26; Giraldo and Trinkunas, 2007b; Naylor, 1993.

[5] Naylor, 1993.

such as al-Qa'ida and its affiliates.[6] Second, the financing units of these groups often do not trust militant operatives, leading to the common problem of underfunded operations.[7]

These findings indicate that facilitating collective action is a key problem for militant groups. They can do so through providing personal benefits, by threatening punishments, or by promising that their success will lead to future benefits for their members. However, providing direct financial incentives to group members will tend to attract individuals whose allegiance can be bought or who might steal from the group.[8] To counteract this limitation, militant groups try to select loyal members and cultivate allegiance to the group. Religious sects and religious militant groups deter casual members from joining by requiring that inductees engage in seemingly irrational behaviors, such as sacrifices and prohibitions. Sacrifices, such as violent initiation rites, may reduce the value of the inductee in legitimate society and therefore reduce the lowest wage at which he would accept legitimate employment, known as the "reservation wage," thus decreasing the perceived benefits of defecting in the future.[9] In exchange for these sacrifices, the group provides "club goods" or selective incentives, such as salaries and martyr payments, to their members.[10]

---

[6] Jacob N. Shapiro, "Terrorist Organizations' Inefficiencies and Vulnerabilities: A Rational Choice Perspective," in Giraldo and Trinkunas, 2007a.

[7] Jacob N. Shapiro and David A. Siegel, "Underfunding in Terrorist Organizations," paper presented at a meeting of the American Political Science Association, Omni Shoreham, Washington Hilton, Washington, D.C., September 1, 2005b.

[8] Eli Berman, "Hamas, Taliban and the Jewish Underground: An Economist's View of the Radical Religious Militias," National Bureau of Economic Research Working Paper W10004, Cambridge, Mass., 2003; Shapiro and Siegel, 2005b.

[9] Laurence R. Iannaccone, "Sacrifice and Stigma: Reducing Free-Riding in Cults, Communes, and Other Collectives," *The Journal of Political Economy*, Vol. 100, No. 2, 1992, pp. 271–291; Eli Berman and David D. Laitin, "Religion, Terrorism and Public Goods: Testing the Club Model," *Journal of Public Economics*, Vol. 92, Nos. 10–11, 2008, pp. 1942–1967; Berman, 2003.

[10] Club goods are goods or services that are nonrivalrous but excludable in consumption. In other words, nonmembers of a club can be denied the benefits of the good, but consumption of the good by one member does not prevent consumption by other members.

Although club goods directly benefit only individuals associated with a militant organization, the organization may also provide public goods to the broader society.[11] Some scholars argue that militant groups provide club and public goods to burnish their public image and project the capability of an alternative sovereign state, and in doing so, they make association with and support of their organization more desirable.[12] However, social service provision by the legitimate government or partner nation can attract greater cooperation from and information sharing by the society at large.[13]

Individual Returns to Militancy

A number of economists have empirically analyzed personal returns to militancy. In a rational-choice approach, individuals who decide to join a militant group must receive greater utility in doing so than by pursuing their next best option. Individuals can receive two types of utility from militant group membership: utility from their compensation and utility from their perceived contribution toward the group's goals. Individuals may weight these two components differently. In a simple mathematical representation of this concept, constraining the two weights to add to one provides a scale from zero to one indicat-

---

[11] Public goods are nonrivalrous and nonexcludable in consumption and so, unlike club goods, cannot be limited to a particular group.

[12] Mark Harrison, "An Economist Looks at Suicide Terrorism," *World Economics*, Vol. 7, No. 3, 2006, pp. 1–15; Justin Magouirk, "The Nefarious Helping Hand: Anti-Corruption Campaigns, Social Service Provision, and Terrorism," *Terrorism and Political Violence*, Vol. 20, No. 3, 2008, pp. 356–375.

[13] The government can gain popular support through the provision of public goods in a number of ways. These include (1) showing the population that the provision of social services signals that a long-term redistribution of wealth will happen if the government wins; (2) proving to the population the government's capacity to provide services and win the war, thus incentivizing bandwagoning; and (3) providing public services to reduce the community's dependence on rebel groups and increase its dependence on the government, leading to stronger relationships and information flow. See Eli Berman, Jacob N. Shapiro, and Joseph Felter, "Can Hearts and Minds Be Bought? The Economics of Counterinsurgency in Iraq," National Bureau of Economic Research Working Paper W14606, Cambridge, Mass., 2008.

ing completely ideological to completely motivated by compensation; the distribution of the summed weights for all members of a group describes the distribution of motivations.[14]

No studies have directly measured and compared militant incomes to their nonmilitant counterparts, but a significant body of research exists on economic correlates of participation in militancy. Research focusing specifically on terrorists has found that more years of education and higher incomes have been associated with participation in terrorism and supporting terrorism; those factors have also been found to be associated with greater effectiveness at carrying out acts of terrorism, suggesting that there are returns to human capital in the conduct of terrorist operations.[15] The processes and incentives involved in militant recruitment are also obscure. Although there are no economic studies observing the recruitment process itself, the observed higher-than-expected education and wealth levels of terrorists weakens theories explaining individual participation in militancy as being due to financial deprivation, mental instability, or poor education.[16]

---

[14] Shapiro, 2007a.

[15] Efraim Benmelech and Claude Berrebi, "Human Capital and the Productivity of Suicide Bombers," *The Journal of Economic Perspectives,* Vol. 21, No. 3, Summer 2007b, pp. 223–238; Claude Berrebi, "Evidence About the Link Between Education, Poverty, and Terrorism Among Palestinians," *Peace Economics, Peace Science, and Public Policy,* Vol. 13, No. 1, 2007, pp. 1–36; Christine C. Fair and Bryan Shepherd, "Who Supports Terrorism? Evidence from Fourteen Muslim Countries," *Studies in Conflict and Terrorism,* Vol. 29, No. 1, 2006, pp. 51–74.

[16] Berrebi, 2007; Efraim Benmelech and Claude Berrebi, "Attack Assignments in Terror Organizations and the Productivity of Suicide Bombers," National Bureau of Economic Research Working Paper W12910, Cambridge, Mass., 2007a; Alan B. Krueger and Jitka Malečkova, "Education, Poverty and Terrorism: Is There a Causal Connection?" *The Journal of Economic Perspectives,* Vol. 17, No. 4, 2003, pp. 119–144; Charles Russell and Bowman Miller, "Profile of a Terrorist," in Lawrence Zelic Freedman and Yonah Alexander, eds., *Perspectives on Terrorism,* Wilmington, Del.: Scholarly Resources Inc., 1983, pp. 45–60; Rex A. Hudson, *The Sociology and Psychology of Terrorism: Who Becomes a Terrorist and Why?* A Report Prepared Under an Interagency Agreement by the Federal Research Division, Library of Congress, Washington, D.C., September 1999.

Empirical Work on Militant Finance and the Financing of al-Qa'ida

Despite the extent of research on terrorist financing and the individual returns to terrorism, a limited amount of truly empirical work exists on contemporary militant group financing. Little is known about the precise amounts of money involved in militant group financing and how these funds are managed and moved within a militant organization.[17] Further, militant salaries have not been systematically analyzed alongside nonmilitant incomes within a society. Such analysis would be similar to that of Levitt and Venkatesh in their landmark study of the economics of a drug-selling gang in Chicago.[18] In general, little attention has been given to militant "career paths," the individual economic returns and risk involved in conducting militant violence, or how economic activities (such as the relationship between financing and the group's output—attacks) are organized in light of the fact that the group has no ability to enforce contracts through legal means.

Quantitative data on the organizational level of militant financing is often used only anecdotally, largely because the underground nature of these activities naturally confounds systematic data collection. Few studies to date have analyzed systematic financial data on militant groups; one treated the Basque separatist group *Euskadi Ta Askatasuna*, and two studies analyzed documents maintained by the Palestine Liberation Organization that showed funds being provided to militant elements.[19]

---

[17] Niko Passas, "Terrorism Financing Mechanisms and Policy Dilemmas," in Giraldo and Trinkunas, 2007a.

[18] Steven Levitt and Sudhir Alladi Venkatesh, "An Economic Analysis of a Drug Selling Gang's Finances," *The Quarterly Journal of Economics*, Vol. 115, No. 3, August 2000, pp. 755–789.

[19] Florencio Irabarren, "ETA: Estrategia Organizative y Actuaciones, 1978–1992," Bilbao, Spain: Universidad del Pais Vasco, 1998, pp. 136–152; Israeli Defense Forces/Military Intelligence, *International Financial Aid to the Palestinian Authority Redirected to Terrorist Elements,* June 2002; Rachel Ehrenfeld, *Where Does the Money Go? A Study of the Palestinian Authority,* New York: American Center for Democracy, 2002.

More closely related to this monograph, few specific data have come to light about the organization of al-Qa'ida–affiliated groups that would allow researchers to draw conclusions about an administrative unit's relationship with the working cadre. The only such publication on an al-Qa'ida–affiliated group assessed financial data on al-Qa'ida in Iraq. The study showed a bureaucratic and highly organized group that sustained itself largely by receiving funds from its foreign recruits, but this analysis was limited, because the data covered only the affairs of a small logistical unit charged with processing foreign fighters arriving to volunteer at Iraq's northwestern border.[20]

Aside from that study, the information that has surfaced has not been analyzed in depth. This is especially true of these organizations' financial administration. Besides the work on foreign fighters destined for AQI, most of the previous organizational analysis was drawn from captured documents specific to al-Qa'ida's central command and its close allies before September 11, 2001, rather than to more recent affiliates. This includes documents in the Harmony database maintained by the Combating Terrorism Center at West Point and other documents the *Wall Street Journal* found on a computer from Ayman al-Zawahiri's office. The analysis of Zawahiri's documents offers only scattered anecdotes of strained interpersonal relationships and budgeting problems.[21]

One overriding theme that does emerge from previous analysis is the core al-Qa'ida's aspiration to establish and maintain a hierarchy, even a bureaucracy, generally ruled from the top down, as a probable precursor to a global caliphate.[22] In this same vein, the Harmony al-Qa'ida documents include bylaws, committee structures, letterheads, forms, and permission slips.[23] The financial data from Iraq reflect the same penchant, although it is unclear whether AQI founder Abu Mus'ab al-Zarqawi imitated al-Qa'ida "central" or merely shared

---

[20] Jacob Shapiro, "Bureaucratic Terrorists: Al-Qa'ida in Iraq's Management and Finances," in Brian Fishman, ed., *Bombers, Bank Accounts and Bleedout: Al-Qa'ida's Road In and Out of Iraq*, U.S. Military Academy, West Point, N.Y.: Combating Terrorism Center, 2008.

[21] Alan Cullison, "Inside Al-Qaeda's Hard Drive," *The Atlantic Monthly*, September 2004.

[22] Security Service (United Kingdom [UK] MI5), "Al Qaida's Ideology," undated.

[23] See Combating Terrorism Center, Harmony Project Database, West Point, N.Y., undated.

the same proclivities. In either case, the data underscore that AQI is a hierarchical organization, not just a loose network of cells. In fact, the financial administration has its own emirs, or commanders, and their geographic portfolios reflect AQI's ambition to govern territory.

The Contribution of This Monograph

In this monograph, we directly analyze a wide array of economic issues related to militancy by using a unique data set containing detailed financial information on al-Qa'ida in Iraq. The data span a 20-month period at the height of the group's influence. AQI's administrative emir of Anbar province maintained these data as a management tool for both tracking and administering the finances and resources of the group in the province. Updated biweekly, the data include daily revenues and expenditures for the provincial administration, payroll sheets for the individual battalions and brigades within specific geographic areas, and memos and reports from the various bureaucratic units of the group. We supplemented these financial data with information on the group from U.S. military analysts and operators in Anbar province at the time, and with data on attacks in Anbar province from the Department of Defense's Significant Activities (SIGACTS) III database.

Although the data have important limitations and possible biases, they allow us to analyze the individual and collective actions of AQI within the structure of a bureaucratic organization that appears to act rationally and respond to incentives. First, we assess the returns to participating in militancy relative to other labor market activities. The higher the relative returns to militancy, the more likely it is that economic aspects of group membership are paramount. We then analyze the trade-offs of militants based on the risks they face and whether these can be considered to be optimal decisionmaking strictly in terms of financial rewards. Finally, we examine the statistical associations between the group's expenditures and various attack measures in Anbar province.

How This Monograph Is Organized

The monograph has six chapters. Chapter Two describes the setting of Anbar province, provides a brief history of AQI, and reviews the group's economic and political environment in Anbar province at the time covered by our study. Chapter Three describes the AQI financial data set, traces the observed money flows across the group's organizational units, and describes the group's finances and expenses throughout the same period. Chapter Four presents an economic analysis of AQI compensation, examining the relative economic benefits of being an AQI member. Chapter Five continues the economic analysis of AQI by analyzing the associations between the group's money flows and the observed attack levels during the period. Chapter Six gives conclusions and policy recommendations.

CHAPTER TWO

AQI and the Political and Economic Environment in Anbar Province

Our data span the period when AQI reached the zenith of its power, as well as the period when the political situation rapidly turned against the group and dramatic changes took place in the security environment in Anbar. This chapter provides a thorough description of AQI and its political and economic environment to put our analysis and assessments of the group's financial data in broader context.

We provide a brief description of Anbar province to place the study in physical space. We then provide a brief history of AQI to describe the group's goals, its membership, and its relationship with both al-Qa'ida central and the Iraqi populace. We next review the political and security situation in Anbar over the period and explain AQI's tumultuous relationship with Anbar's tribes. Appendix B contains a time line of key events during the period. Finally, we describe the economic situation in Anbar during the period under consideration, early 2005 through the end of 2006, from available economic surveys.

Anbar Province

Anbar is geographically the largest province in Iraq and has a majority Sunni population of over 1.4 million.[1] The province lies in the western region of the country, bordering Saudi Arabia, Syria, and Jordan.

---

[1] Central Organization for Statistics and Information Technology (COSIT), Ministry of Planning and Development Cooperation, Iraq; Kurdistan Region Statistics Office (KRSO), Iraq; Nutrition Research Institute, Ministry of Health, Iraq; and United Nations World

The desert environment causes extreme temperature changes and a dearth of water sources, and, as a result, most of the inhabitants live near the Euphrates River valley, which stretches from al-Qa'im in the northwest through Fallujah in the east.[2] Anbar officially has eight distinct geographic districts. From west to east along the Euphrates River valley, these include al-Qa'im, Anah, Rawah, Haditha, Hit, Ramadi, and Fallujah; Rutbah, the eighth, is largely desert and covers the entire southern part of the province from the western to the eastern borders, although it does not extend as far east as Fallujah.[3] Appendix A provides a detailed map of the province (Figure A.1) and a map with our assessment of AQI sectors in the province (Figure A.2).

AQI: A Primer

After the United States and its partners, the Coalition Forces (CF), invaded Iraq in March 2003 and deposed the Saddam Hussein government in April, a nascent Sunni insurgency coalesced around an amalgam of former Baathists, nationalists, salafi jihadists, and criminals.[4] Some commentators have described the early period of the insurgency in 2004 and 2005 as a "composite insurgency" that had no discernible leadership.[5] The multifarious militant groups rallied around the single

---

Food Programme, Iraq Country Office, *Comprehensive Food Security & Vulnerability Analysis in Iraq 2007*, 2008.

2 Inter-Agency Information and Analysis Unit, "Anbar Governorate Profile," United Nations Office for the Coordination of Humanitarian Affairs, Amman and Baghdad, March 2009.

3 COSIT, 2008.

4 Salafi characterizes an adherent of an ideological strain in Sunni Islam that seeks to emulate, as purer, the thinking and practices of Muhammad and the earliest generations of Muslims. Jihadists believe that violent struggle against non-Muslims and Muslims they judge as apostate is an important religious duty. AQI can be considered a salafi jihadist group.

5 Michael Eisenstadt and Jeffrey White, "Assessing Iraq's Sunni Arab Insurgency," Washington Institute for Near East Policy, Policy Focus #50, December 2005; Seth Jones and Martin Libicki, *How Terrorist Groups End: Lessons for Countering al Qa'ida*, Santa Monica, Calif.: RAND Corporation, MG-741-1-RC, 2008.

cause of resisting the U.S. occupation. They built their interactions on a loose network of deep-seated familial, tribal, and local loyalties. However, over the course of the first two years, the Saddam loyalists rapidly lost influence. Through the course of 2004 and 2005, even the nationalists shifted to a salafi jihadist discourse. The larger movement came to be dominated by a few large salafi groups, namely AQI, Jama'at Ansar al-Sunnah, the Islamic Army in Iraq, and the Islamic Front for Iraqi Resistance.[6]

As early as mid-2003, Jordanian extremist Abu Mus'ab al-Zarqawi attempted to raise the profile of AQI, known at the time as Jama'at al-Tawhid wa al-Jihad (JTJ, or the Monotheism and Jihad Group) among salafi jihadists in Anbar. He used a variety of tactics, including publicizing the group's horrific kidnappings and beheadings.[7] Before late 2001, Zarqawi had run training facilities for foreign jihadists in Afghanistan. JTJ attracted mainly foreign volunteers in the first year of the Iraqi insurgency.[8] Although JTJ gradually incorporated more Iraqis to downplay its foreign identity, foreign fighters continued to make up more than half of suicide attackers in Iraq working for Zarqawi and his successor into 2007, according to estimates by the Combating Terrorism Center at West Point in a study of foreign fighter documents recovered from Sinjar, a western Iraqi town near the Syrian border.[9] JTJ centered its activities in Anbar around Fallujah until Coalition Forces cleared the city in November 2004.

Partly in an effort to further enhance his group's broader pretension of creating a regional caliphate, in October 2004 Zarqawi declared his allegiance to Usama Bin Ladin and merged his group with al-Qa'ida "central" to form Tanzim Qa'idat al-Jihad fi Bilad al-

---

[6] International Crisis Group, "In Their Own Words: Reading the Iraqi Insurgency," Amman and Brussels, 2006.

[7] One prominent example was the brutal beheading of Nicholas Berg in May 2004, which was videotaped and spread across the Internet.

[8] Mary Anne Weaver, "The Short, Violent Life of Abu Musab al-Zarqawi," *The Atlantic*, July/August 2006; Adel Darwish, "Abu Musab al-Zarqawi," *The Independent*, June 9, 2006.

[9] Brian Fishman, ed., *Bombers, Bank Accounts & Bleedout: Al-Qa'ida's Road In and Out of Iraq*, U.S. Military Academy, West Point, N.Y.: Combating Terrorism Center, 2008.

Rafidayn, otherwise known as al-Qa'ida in Iraq.[10] After the loss of Fallujah, the group headquartered its Anbar operations near the provincial capital of Ramadi. The larger set of financial documents discussed in this monograph was captured in Julaybah, a town just east of Ramadi.

Zarqawi and AQI were the targets of criticism within Iraq that they were a foreign element unrepresentative of Sunni or Iraqi national aspirations. A 2005 letter from Bin Ladin's deputy Ayman al-Zawahiri underscored to Zarqawi the need to accommodate the local population to retain the critical public support necessary to establish a territorial foothold for an emirate.[11] In an effort to demonstrate a semblance of unity, in January 2006, AQI declared the formation of a Mujahidin Shura Council (MSC), nominally an umbrella of Iraqi Sunni jihadist groups with AQI a mere member.

From a funding perspective, the merger with al-Qa'ida may have imposed a financial burden on AQI. The *New York Times* reported in 2006 that the U.S. National Security Council believed that the insurgency in Iraq was self-financing.[12] In 2007, the *Los Angeles Times* reported that the U.S. intelligence community believed that AQI's criminal enterprise was so lucrative that it was able to send excess money to al-Qa'ida senior leaders in the Pakistan border region in 2006.[13]

Much of this money may have been coming from refined products stolen from the Bayji oil refinery and then sold on the black market. In early 2007, Iraqi and U.S. officials, respectively, believed that $1 billion or $2 billion of refined product had disappeared into the black market.

---

[10] The group's name is literally translated as "the Base for Jihad in the Land of Two Rivers," but we have used the name "al-Qa'ida in Iraq" throughout this monograph, as it is the most commonly used name for the group in the West. The group is also known as al-Qa'ida in Mesopotamia, and the Organization of al-Qa'ida in the Land of the Two Rivers. In the documents we analyze, the group calls itself the "Islamic State of Iraq"—the name of the umbrella group that AQI created to subsume the broader Sunni insurgency in 2006.

[11] Office of the Director of National Intelligence, "Letter from al-Zawahiri to al-Zarqawi," ODNI News Release No. 2-05, October 11, 2005.

[12] John F. Burns and Kirk Semple, "U.S. Finds Insurgency Has Funds to Sustain Itself," *New York Times*, November 26, 2006.

[13] Greg Miller, "Bin Laden Hunt Finds Al Qaeda Influx in Pakistan," *Los Angeles Times*, May 20, 2007.

U.S. officials futher estimated that insurgents generated as much as $200 million annually from oil smuggling. Insurgent revenues from Bayji-related activities fell to an estimated $50,000 to $100,000 per day by early 2008, according to Iraqi security experts, in large part because a CF base was installed inside the refinery and because of the actions of a new refinery director general and Oil Protection Force colonel.[14]

The al-Qa'ida leaders in Pakistan believed that AQI had excess funding that it could share. In a captured 2005 letter, Zawahiri asked Zarqawi for "one hundred thousand" of unspecified currency, indicating that "we need a payment while new lines are being opened."[15] If funds were in fact transferred to central al-Qa'ida, the documents under review here would not have shown the transactions, because the funds probably would have come from AQI's national general treasury (al-Khazinah or Bayt al-Mal), which the documents show only as a recipient of unused funds from AQI in Anbar province.

Opposition to AQI mounted in 2006 as the group increasingly targeted Sunni tribal leaders and rivals as government collaborators or spies.[16] After Zarqawi's death in June, his successor, Abu Ayyub al-Masri, also known as Abu Hamza al-Muhajir, moved to strengthen AQI's position by laying claim to territory in the guise of a new emirate under an Iraqi leader. He had the MSC declare the establishment of the "Islamic State of Iraq" (ISI) in October 2006 under Abu Umar al-Baghdadi and made himself "minister of war." Although AQI continued a strong pace of attacks using improvised explosive devices (IEDs) and other means in Anbar despite Zarqawi's death, its influence and reputation in Anbar began to wane soon after the formation of the

[14] Ayman Oghanna, "Corruption Stemmed at Beiji Refinery—But for How Long?" *Iraq Oil Report* (posted by Alice Fordham), February 16, 2010; Richard A. Oppel, Jr., "Iraq's Insurgency Runs on Stolen Oil Profits," *New York Times*, March 16, 2008. Even as of early 2010 the problem had not been solved, with diversion of tanker trucks reportedly continuing (Oghanna, 2010).

[15] Office of the Director of National Intelligence, 2005.

[16] Brian Fishman, "Dysfunction and Decline: Lessons Learned from Inside al-Qa'ida in Iraq," Harmony Project, U.S. Military Academy, West Point, N.Y.: Combating Terrorism Center, March 16, 2009.

ISI. First, the declaration of the state drew criticism on religious and political grounds from Sunnis across the Arab world.[17] Next, Sunni tribal leaders formed the "Awakening" movement in opposition to the jihadists. In addition, rival insurgent organizations drew greater public support, according to research published by the Combating Terrorism Center at West Point.[18]

At the same time, a lengthy coalition offensive against the insurgency in Ramadi cleared the city by the spring of 2007, forcing ISI to withdraw to areas mainly outside Anbar after the middle of the year.[19] ISI's financial management may have deteriorated after their ouster from Anbar in 2007. According to a West Point Combating Terrorism Center study, a 2008 captured "lessons learned" document provided an assessment of ISI's strategic and bureaucratic failures that may have contributed to the loss of ground in Anbar. The document's author explained that AQI had become overly bureaucratized, with too many self-appointed emirs and overspecialization leading to stovepiping.[20] On the financial side, the document maintained that ISI lacked a regular funding source and showed favoritism in distributing the money it did receive.[21] As of late 2010, ISI survives principally in northern Iraq, but it retains a capability to conduct spectacular attacks across the country. The group carried out several vehicle-borne explosive strikes against targets across the country in the second half of 2009 and in the first half of 2010. Despite the deaths of al-Masri and al-Baghdadi in a joint U.S. and Iraqi raid in April 2010 in the Tikrit area, the group

---

[17] Evan F. Kohlmann, "State of the Sunni Insurgency in Iraq: August 2007," New York: The NEFA Foundation, 2007.

[18] Fishman, 2009.

[19] Kimberly Kagan, "The Anbar Awakening: Displacing al Qaeda from Its Stronghold in Western Iraq," *Iraq Report*, Washington, D.C.: Institute for the Study of War, August 21, 2006–March 30, 2007.

[20] The Arabic word "emir" means "commander." Although often thought to denote someone of high office or nobility, in AQI there were a large number of emirs who were afforded this status despite only being a leader of a small unit or minor section of the group's bureaucracy.

[21] See Fishman, 2009. Note that this internal memo may conflict with the 2006 *New York Times* report citing AQI's consistent funding sources and financial self-sufficiency. However, the discrepancy may be a result of AQI taking stock of its declining fortunes after 2006.

has been able to continue operations, has named new leaders, and has announced a new campaign against Iraqi security forces.[22]

Politics and Security in Anbar

The political context that AQI operated within in Anbar province included three components: (1) the tribes of Anbar province, (2) the U.S. and Iraqi security forces in Anbar, and (3) the broader Sunni insurgent movement. In this section, we present a brief chronological review of the major shifts in these three elements to show the ebb and flow of AQI's role in society and its ability to influence, and be influenced by, the politics of Anbar province.

Tribes and Security in Modern Iraq

Despite numerous attempts by both the Saddam Hussein government and the Coalition Provisional Authority (CPA) to brush them aside, tribes have been and continue to be important parts of the political fabric of Anbar province and Iraq more broadly.[23] These tribal allegiances, which are based on "fictive kinship" or a myth of common ancestry, regulate both political conflict and economics in areas where the state fails to do so. Tribes also constitute a competing power center for the formal institutions of the Iraqi state and form a parallel hierarchy that overlaps the structures and the political allegiances of formal government at every level.[24] Although tribes are nearly universal in

---

[22] Michael Roddy, "Qaeda Confirms Deaths of Leaders in Iraq: Statement," *Reuters*, April 25, 2010; Aseel Kami and Michael Christie, "Al Qaeda's Iraq Network Replaces Slain Leaders," *Reuters*, May 16, 2010; Liz Sly, "Top Two Al-Qaeda in Iraq Leaders Are Dead, Officials Say," *Los Angeles Times*, April 20, 2010.

[23] The United States and CF set up the Coalition Provisional Authority as the occupation authority following the immediate overthrow of the Saddam Hussein government. The CPA governed from May 2003 to June 28, 2004, when the CPA formally transferred sovereignty to an Iraqi interim government.

[24] David J. Kilcullen, "Field Notes on Iraq's Tribal Revolt Against Al-Qa'ida," *CTC Sentinel,* Vol. 1, Issue 11, October 2008.

Iraq, they are not monolithic; tribes are composed of clans, which in turn are composed of smaller extended families.[25]

The power dynamics between the tribes of Iraq and the Iraqi state are important for understanding how and why AQI rose and fell between 2004 and 2007. Realizing that tribes were integrated in Iraqi society both deeply and broadly, in the 1920s the British relied on the tribes to provide local security during colonization.[26] Following the founding of the modern Iraqi state between the two world wars, the importance of tribes has ebbed and flowed with the strength of the Iraqi state. However, tribes always held a significant amount of political influence. Tribal power reached its nadir in the late 1960s. The Baath Party takeover in 1968 could have ostensibly eliminated the role of the tribes in Iraq because of the Baath's stated goals of modernization and secularization. However, the weak internal position of the regime led Saddam to use his tribal affiliations to assure control of the security institutions of the state. The Iran-Iraq war in the 1980s sent many of these loyal tribesmen to the front lines, forcing Saddam to woo new tribes to garner additional support.

Iraq's 1991 invasion of Kuwait and the resulting international military intervention led by the United States resulted in severe casualties for the Iraqi military, which created an opening for the Shi'a revolt in southern Iraq. The revolt increased Saddam's sense of vulnerability. He again responded by increasing the role of tribes in providing local security. The state's affiliation with and reliance on the tribes became more explicit after 1991, as Saddam increasingly let the tribes independently police local areas and offered increased deference to tribal customs (*'adat*). In 1996, Saddam gave tribal chiefs authority over judicial affairs, gave them official local security powers, and even allowed them to collect taxes for the state. This last provision increased the role of the tribes in smuggling, highway carjackings, and extortion rings. The impending invasion of the United States in 2003 led Saddam to integrate the tribal forces with military and paramilitary formations to

---

[25] Austin Long, "The Anbar Awakening," *Survival,* Vol. 50, No. 2, April–May 2008, pp. 67–94.

[26] This section draws on Long, 2008.

prevent another uprising in the face of a foreign invasion. However, the tribes sensed the weakness of the regime and quickly reneged on supporting the Baath regime after CF invaded in May 2003.

Insurgency, Tribes, and the Security Forces in Anbar Province: 2003–2007

The CF invasion of Iraq was a watershed for the internal politics of Anbar province. The de-Baathification law implemented by the CPA in 2003 had a monumental political and economic effect on Anbar province because of the social and economic displacement of the large concentration of Sunni officers and former regime elements who resided there. The decision to officially disband the Iraqi military also stung the pride of a wide array of Iraqis; the military was the single institution that the entire country took pride in.[27] Further, American commanders have noted that CF heavy-handedness in Fallujah in 2003 and 2004 created specific grievances against the coalition.[28] The broader insurgency took advantage of this social discontent in the summer of 2003. The movement became more radical over time, with AQI taking a lead role, primarily in Fallujah.

CF engagement of the tribes began in early 2004 but with very modest results.[29] AQI's gruesome public killing and dismemberment of four Blackwater guards sparked CF engagement in the first battle of Fallujah in early 2004, called Operation Vigilant Resolve. The U.S. Marines succeeded in clearing most extremist elements in the operation, but the leaders of Multi-National Force–Iraq (MNF-I), the over-

---

[27] John F. Kelly (Lieutenant General, U.S. Marine Corps), "Foreword," pp. vii–x, in Timothy S. McWilliams (Chief Warrant Officer-4, U.S. Marine Corps Reserve) and Kurtis P. Wheeler (Lieutenant Colonel, U.S. Marine Corps Reserve), eds., *Al Anbar Awakening: Volume I, American Perspectives—U.S. Marines and Counterinsurgency in Iraq, 2004–2009,* Marine Corps University, United States Marine Corps, Quantico, Va.: Marine Corps University Press, 2009.

[28] Kelly, 2009.

[29] Long, 2008; interview with James T. Conway (Lieutenant General, U.S. Marine Corps, and Commanding General, I Marine Expeditionary Force [I MEF], November 2002 to September 2004), in McWilliams and Wheeler, 2009, pp. 40–58; interview with Michael M. Walker (Colonel, U.S. Marine Corps, and Commanding Officer, 3d Civil Affairs Group, I MEF, October 2003 to September 2005), in McWilliams and Wheeler, 2009, pp. 60–74.

all CF command in Iraq, called for an early withdrawal from the city because of the public relations disaster created by the heavy use of firepower. MNF-I later realized its mistake in allowing the remaining terrorists to retain a safe haven there and embarked on Operation Al Fajr, the second battle of Fallujah, in late 2004 to clear and hold the city. Al Fajr successfully cleared Fallujah of extremist insurgents and set up police forces backed by local tribes, and civil affairs teams began a lasting reconstruction of the city with local buy-in.[30]

Despite the gains made in restive Fallujah, the MNF-I command's strategic review at the end of 2004 concluded that the insurgency was fundamentally a reaction to the presence of foreign troops in the streets and the politics of Iraq and decided that their best efforts should be put toward more quickly training the lagging Iraqi Army (IA) so that the Iraqis could stand up as CF stood down.[31] The resulting force posture of the coalition became more relaxed, with many patrol forces pulling back to forward operating bases to train the Iraqi Army, increasing insecurity and reducing intelligence collection.[32] The strategy failed for two reasons. First, it compromised security in the short run. Second, the new Iraqi Army could not establish the trust of the Anbari populace because it was seen as a Shi'a-dominated force.[33]

The clearing of Fallujah, meanwhile, started a shift in the attitudes of the tribes at the beginning of 2005. When many of the Sunni tribes boycotted the provincial elections in January, the security vacuum allowed AQI to compete for tribal revenue sources, and AQI's increasingly transnational focus was more and more clearly at odds with the

[30] Long, 2008; interview with Richard F. Natonski (Major General, U.S. Marine Corps, Commanding General, 1st Marine Division, Multi-National Force–West [MNF-W], August 2004 to March 2005) in McWilliams and Wheeler, 2009, pp. 88–96; Carter Malkasian, "A Thin Blue Line in the Sand," DemocracyJournal.org, Summer 2007.

[31] Interview with Natonski in McWilliams and Wheeler, 2009, pp. 88–96; Malkasian, 2007.

[32] Malkasian, 2007; interview with Alfred B. Connable (Major, U.S. Marine Corps, Senior Intelligence Analyst/Fusion Officer, I and II Marine Expeditionary Forces, 2005–2006), in McWilliams and Wheeler, 2009, pp. 120–137.

[33] Interview with Connable in McWilliams and Wheeler, 2009, pp. 120–137; Malkasian, 2007.

local and nationalist goals of the tribes.[34] The Albu Mahal tribal leaders in western Anbar contacted MNF-W, the CF's regional command in Anbar, for military support in al-Qa'im against AQI, but a previously planned clearing operation called Operation Matador used heavy firepower and destroyed parts of the city, alienating the Albu Mahal.[35] However, the Dulaimi tribal federation continued to fight AQI in the area, taking heavy losses in Ramadi through the early fall of 2005. Although MNF-W was slow to embrace the shifting allegiances of the tribes, by November, the command planned a coordinated security operation called Operation Steel Curtain with the Albu Mahal and successfully cleared al-Qa'im.[36] Both the IA and CF stayed behind to provide lasting security pressure in the area.[37]

In December 2005, a pause in violence occurred as tribal leaders came to realize the benefits of participating in electoral politics and subsequently endorsed the national elections. A group of nationalist tribal sheiks established the Anbar People's Council to provide security and fight AQI, and in January 2006 there was a surge in recruitment for the security forces. AQI responded by conducting a series of suicide bombings and assassinations that had a dramatic chilling effect on the political climate. By the end of January, more than half of the Anbar People's Council leaders had been assassinated. The level of violence in the center of the province rose dramatically from 25 incidents a day to 90 a day. AQI successfully convinced the nationalists, who had joined the insurgency in its early period, that their survival depended on joining AQI, and began absorbing these groups into the AQI ranks.[38]

---

[34] Long, 2008. A paltry 3,700 votes were cast in the January 2005 provincial elections in Anbar province versus approximately 500,000 in the December 2005 national elections.

[35] Long, 2008.

[36] Interview with Stephen T. Johnson (Major General, U.S. Marine Corps, Commanding General, II Marine Expeditionary Force [Forward], MNF-W, February 2005 to February 2006), in McWilliams and Wheeler, 2009, pp. 100–110; interview with James L. Williams (Brigadier General, U.S. Marine Corps, Assistant Commanding General, 2d Marine Division, July 2005 to January 2006), in McWilliams and Wheeler, 2009, pp. 112–118; interview with Connable, in McWilliams and Wheeler, 2009, pp. 120–137.

[37] Long, 2008.

[38] Interview with Connable, in McWilliams and Wheeler, 2009, pp. 120–137.

At the outset of 2006, Anbar was largely cleared of extremists in the far west and east, but the entire center of the province and the provincial capital of Ramadi remained extremely violent; in fact, it was the most dangerous place in Iraq at the time.[39] Although CF security operations and targeted raids increased through 2006, security incidents continued to rise. General Robert B. Neller, a commander in Anbar under MNF-W, explained this as a result of CF moving into previously denied areas, but Colonel Pete Devlin and Major Ben Connable of Marine Corps intelligence attributed this more to AQI's exercise of its political power in the face of tribal resistance and security force pressure, as AQI was at the time the "dominant organization of influence in Anbar province, surpassing the nationalist insurgents, the Government of Iraq, and MNF-I in its ability to control the day-to-day life of the average Sunni."[40]

In March and April 2006 General Richard C. Zilmer of MNF-W drew up plans with the IA to secure Ramadi without the support of the tribes. Security stations and checkpoints were meticulously laid out.[41] More security forces arrived in the summer of 2006. After AQI's bombing of a hotel in Amman in late 2005, the bombing of a major Shi'a shrine in Samarra and subsequent sectarian violence in early 2006, assassinations of tribal leaders, and intense fighting between AQI and

[39] Interview with Connable, in McWilliams and Wheeler, 2009, pp. 120–137; interview with Richard C. Zilmer (Major General, U.S. Marine Corps, Commanding General, I MEF [Forward], MNF-W, February 2006 to February 2007), in McWilliams and Wheeler, 2009, pp. 140–151; interview with David G. Reist (Brigadier General, U.S. Marine Corps, Deputy Commanding General [Support], I MEF [Forward], MNF-W, February 2006 to February 2007), in McWilliams and Wheeler, 2009, pp. 152–161; interview with Robert B. Neller (Brigadier General, U.S. Marine Corps, Deputy Commanding General [Operations], I MEF [Forward], MNF-W, February 2006 to February 2007), in McWilliams and Wheeler, 2009, pp. 162–174; interview with Sean B. McFarland (Colonel, U.S. Army, Commanding Officer, 1st Brigade Combat Team, 1st Armored Division, U.S. Army, Multi-National Force–North, January 2006 to June 2006, MNF-W, June 2006 to February 2007), in McWilliams and Wheeler, 2009, pp. 176–185.

[40] Interview with Connable, in McWilliams and Wheeler, 2009, pp. 120–137.

[41] Interview with Connable, in McWilliams and Wheeler, 2009, pp. 120–137.

CF in Ramadi, both the central Anbari populace and the tribal sheiks began to support CF.[42]

Scholars and military commanders who were present in Anbar during the period emphasize that the tribal sheiks finally turned to support CF in the late summer of 2006 out of self-interest rather than moralistic considerations. Chief among these interests was self-preservation in the face of AQI murder and intimidation designed to control and weaken the tribal elders, as well as AQI infringement on tribal smuggling rackets, extortion practices, and highway theft.[43] In September 2006, a coalition of 17 tribal sheiks led by Sheik Sattar Abu Risha announced the formation of the Sahwat al-Anbar (SAA, or "the Anbar Awakening"), a tribal movement aimed at neutralizing AQI by providing local security by creating a new militia force and encouraging increased recruitment into the Iraqi police. Sattar was a smuggler, a highway robber, and a relatively minor sheik who first tried to align the tribes against AQI in 2005 as the group was infringing on his business interests.[44]

Unlike the prior tribal revolt in western Anbar, this movement took hold and spread across Anbar and eventually the rest of Iraq.[45] This can be explained primarily by the fact that the government of Iraq (GoI) and Prime Minister Nouri al-Maliki officially recognized the SAA movement, and the GoI officially gave tribal leaders increased political, military, and economic power. In particular, the GoI named Sattar counterinsurgency (COIN) coordinator for the province, deemed his militia an official "emergency response unit," and gave him control of three local IA battalions.[46] Concurrently, the GoI gave the

---

[42] Interview with Connable, in McWilliams and Wheeler, 2009, pp. 120–137; Kelly, 2009.

[43] Long, 2008; Kelly, 2009; interview with Connable in McWilliams and Wheeler, 2009, pp. 120–137.

[44] Long, 2008.

[45] Interview with William M. Jurney (Lieutenant Colonel, U.S. Marine Corps, Commanding Officer, 1st Battalion, 6th Marines, assigned to 1st Brigade Combat Team, 1st Armored Division, U.S. Army, I MEF [Forward], September 2006 to May 2007), in McWilliams and Wheeler, 2009, pp. 186–199.

[46] Long, 2008.

Albu Mahal tribe control over a resident army brigade, the authority to fill the brigade's ranks with Albu Mahal tribesmen, and the freedom to retake control over smuggling routes into Syria.[47] In all, this new form of tribal cooperation with the security forces swept across Iraq in 2007 under various monikers.[48] The success of the tribal engagement program can be attributed to the fact that it officially linked tribes with formal government structures and was both more public and more dramatic than previous efforts.[49]

In the first four months of 2007, the number of U.S. forces in Ramadi grew when a number of units had their tours extended in anticipation of the arrival of new units from the recently announced "surge," otherwise known as the Baghdad Security Plan.[50] With the addition of new CF units, the recruitment of 4,500 new policemen by Ramadi tribal sheiks, and a series of CF security operations around Ramadi, stability began to take hold in central Anbar. Despite a series of attempted counterattacks by AQI in February and March, including chlorine gas attacks, incidents against CF in Anbar dropped from between approximately 300 to 400 per week to nearly zero per week by the beginning of the spring of 2007.[51]

[47] Malkasian, 2007.

[48] The tribal irregular defense forces were originally called "the Anbar Awakening," but the GoI and MNF-I later renamed the program "Concerned Local Citizens" and finally "Sons of Iraq" as it was expanded across disparate parts of Iraq as a way to avoid the distinctly Sunni and Anbari identity of the SAA political movement.

[49] Long, 2008.

[50] Kagan, 2007.

[51] Kagan, 2007; interview with Walter E. Gaskin, Sr. (Major General, U.S. Marine Corps, Commanding General, II Marine Expeditionary Force [Forward], MNF-W, February 2007 to February 2008) in McWilliams and Wheeler, 2009, pp. 214–224; interview with John R. Allen (Major General, U.S. Marine Corps, Deputy Commanding General, II Marine Expeditionary Force [Forward], MNF-W, January 2007 to February 2008) in McWilliams and Wheeler, 2009, pp. 226–237.

The Economy of Anbar

Economic considerations weighed heavily on tribal elders who chose to take sides in the insurgency. This section sheds light on the welfare of the average Anbari caught between the violence of AQI and the security forces and the constantly shifting allegiances of the local tribes.

We base this section on three data sources: the Iraq Living Conditions Survey, conducted by the United Nations Development Programme, Iraq's COSIT, and Fafo Institute for Applied International Studies of Norway in 2004 and published in 2005; the Iraq Household Socioeconomic Survey, conducted by COSIT, the Kurdistan Region Statistics Organization, and the World Bank in 2007 and published in 2008; and a survey of Anbar province conducted by the RAND Corporation in 2008 and published in 2009.[52] Unfortunately, the extreme insecurity in Anbar precluded large-scale surveys during the period covered by our AQI data, 2005 and 2006.

Agriculture has historically dominated the Anbar economy. As of 2004, 25 percent of its workers were employed in agriculture, hunting, or fishing, compared to only 15 percent of Iraqi workers overall.[53] The people of Anbar province had higher household incomes (in nominal terms) and lower unemployment rates than the people in Iraq overall in 2004. Median biweekly per capita household income was 125,197 Iraqi dinars in Anbar (roughly U.S. $2,300 annually at 2004 exchange rates) and 105,500 Iraqi dinars, or $1,900 annually, for Iraq overall.[54] The unemployment rate was 8.4 percent in Anbar and 10.5 percent

---

[52] Ministry of Planning and Development Cooperation, *Iraq Living Conditions Survey 2004,* Baghdad, Iraq, 2005; Central Organization for Statistics and Information Technology (COSIT), Iraq, KRSO, and International Bank for Reconstruction and Development/The World Bank, *Iraq Household Socio-Economic Survey IHSES-2007,* Baghdad, Iraq, 2008; Keith Crane, Martin C. Libicki, Audra K. Grant, James B. Bruce, Omar Al-Shahery, Alireza Nader, and Suzanne Perry, *Living Conditions in Anbar Province in June 2008,* Santa Monica, Calif.: RAND Corporation, TR-715-MCIA, 2009.

[53] Ministry of Planning and Development Cooperation, 2005.

[54] Ministry of Planning and Development Cooperation, 2005. We converted the Iraqi dinar at a rate of 1,440 dinars per U.S. dollar, the rate used in the publication.

for Iraq.[55] Unemployment was concentrated among young males, but young Anbaris still had lower unemployment rates than Iraqi youth overall. However, labor force participation was slightly higher in Iraq than it was in Anbar province. Anbar's labor market differed markedly from that in the rest of Iraq, with a greater percentage of its population employed in agriculture and a slightly larger percentage of people working in public administration and defense. Iraq overall had a larger percentage of people employed in trade or manufacturing. Family businesses also employed a much higher percentage of people in Anbar than in Iraq; one-quarter of Anbaris worked in a family business, whereas only 11 percent of Iraqis did. Correspondingly, 44 percent of Iraqis were employed in private companies in 2004, but only 28 percent of employed people in Anbar province worked for such companies.[56]

By 2008, the average income of employed individuals in Anbar was just under $84 per week. Unemployment was still concentrated among young males, but no men over age 23 reported being unemployed.[57] As in its 2004 survey, COSIT measured unemployment in 2007 as lower in Anbar than in Iraq overall. By age category, levels of labor force participation were higher for males in Anbar in 2008 than the overall rate for Iraqi males in 2004, except for males in the 15–19 age group; however, they were comparable with overall Iraqi male labor force participation in 2007.[58] The unemployment rate was 7.9 percent in Anbar in 2008 and 11.7 percent for Iraq in 2007.[59] Labor force participation among men age 15 and older had also increased since 2004, from 66 to 75 percent.[60]

---

[55] Ministry of Planning and Development Cooperation, 2005.

[56] Ministry of Planning and Development Cooperation, 2005.

[57] Crane et al., 2009, p. 26. We converted the actual figure in the text, 100,000 Iraqi dinars, at an exchange rate of 1,193, the average exchange rate in 2008 according to Central Bank of Iraq, 2010.

[58] Crane et al., 2009; COSIT, KRSO, and the World Bank, 2008. We note that the definitions of labor force participation in the two documents may not be the same and so comparisons are inexact.

[59] Crane et al., 2009; COSIT, KRSO, and the World Bank, 2008.

[60] Crane et al., 2009.

Drawing on differences between the 2004 and 2008 data and on retrospective questions in the 2008 surveys, the period starting at the end of 2006 when the security situation began to improve also appeared to be a time of economic growth and change. The RAND surveys from May 2008 showed that Anbaris had seen a 9 percent increase in income since December 2006, with wider differences among older Anbaris than among younger. Less than 15 percent of Anbaris saw a decrease in income during that period, and almost half saw some increase.[61] All individuals who reported having had a job in 2006 were also employed in 2008. The economy had rapidly dollarized, likely because of high inflation, with 30 percent of income-earners reporting being paid in U.S. dollars in May 2008, and no one reporting being paid in U.S. dollars in December 2006. Another sign of shift in the Anbar economy is that whereas a greater percentage of Anbaris than Iraqis had been employed in public administration and defense in 2004, the reverse was true in 2008.

[61] Crane et al., 2009.

CHAPTER THREE

Auditing al-Qa'ida in Iraq

Our analysis is based on two novel data sets relating to the administration of AQI in Anbar province in 2006. The first set of documents is referred to as the "Travelstar" documents. Awakening forces found these when they raided the residence of Ala Daham Hanush in Julaybah, Iraq, on March 15, 2007 (see Figure A.1 in Appendix A).[1] The hard drive of Hanush's computer contained the documentation of AQI's provincial administration of Anbar province from 2005 through the end of 2006. This unit of the organization reviewed and regulated the activities of AQI in its different geographic sectors within Anbar province and provided money to these sectors when they needed financing. The master financial ledgers formed the basis of our analysis; these are aggregates of many of the 1,200 files in the collection.[2]

A Marine unit on routine patrol found the second set of nine documents in a ditch in the town of Tuzliyah in western Anbar in January 2007. This set contained AQI financial records, including payrolls, records of equipment and supply purchases, and the flow of funds into and out of AQI's smaller "western" sector of Anbar province from September through mid-December 2006.[3] This western sector was the administrative overseer for AQI's military and political activities in the

[1] Harmony Batch ALA DAHAM HANUSH, documents NMEC-2007-633541, NMEC-2007-633700, NMEC-2007-633893, and NMEC-2007-633919.

[2] The master financial ledgers are NMEC-2007-633541, NMEC-2007-633700, NMEC-2007-633893, and NMEC-2007-633919.

[3] Harmony Batch MA7029-5, documents MNFA-2007-000560, MNFA-2007-000562, MNFA-2007-000564, MNFA-2007-000566, MNFA-2007-000570, MNFA-2007-000572,

western region and reported to the provincial administration of Anbar. Both document sets show interaction between these two distinct levels of AQI command, including financial transfers, memos, and military and administrative reporting.

The analysis of the financial data proceeds in two parts. In this chapter, we present simple descriptions of the money flows, both in terms of who gave and who received funds, as well as how much money was generated and transferred between organizational units across time. The second part of our analysis, in Chapters Four and Five, seeks a deeper economic understanding of the group's financial operations and the incentives that it provided to its members by investigating compensation patterns, the risk of death, and the relationship between spending and attacks.

Area of Operations

Awakening forces seized the Travelstar document set in Julaybah, Iraq, along Highway 10 in March 2007 (see Appendix A). The documents show an AQI "provincial" organization that had purview across all of Anbar province from Rutbah in the west to al-Qa'im in the northwest and Fallujah in the east. Most of the entries pertain to the most highly populated areas in Anbar province proper, that is, the Ramadi-Fallujah corridor and the Haditha area. It is also notable that specific AQI groups in western Anbar, such as those in the sparsely populated Rutbah and al-Qa'im areas, were near key border crossings and were on important trade routes from Syria and Jordan to Baghdad. Finally, there are observed financial flows to two border emirs whose locations are not specified, but whom we believe to be at least partially documented in the Sinjar records analyzed by the Combating Terrorism Center (CTC) at West Point.[4]

MNFA-2007-000573, and MNFA-2007-000574. The Arabic term used for western is "Gharbiyah."

[4] Fishman, 2008.

AQI in Anbar's Organization, Bureaucracy, and Division of Labor

AQI's financial documents depict a hierarchical organization with a centrally controlled bureaucracy that directly funded and influenced the operations of the group's subsidiary "sectors" of Anbar. Using communications data to model terrorist groups as social networks fails to capture the hierarchical aspect of relationships, which may partially explain why hierarchy has, except for Shapiro (2008) in the CTC–West Point study of the Sinjar documents, largely gone unnoticed in the case of AQI and of contemporary terrorism more generally.[5] In these documents, a provincial general emir and an administrative emir served as key decisionmakers for the allocation of finances between the sectors, directing funds to and collecting revenues from each.

AQI's Division of the Anbar Operating Area

AQI divided Anbar into six geographic sectors, each of which had a sector administrator who divided his area into distinct subunits or subsectors. Each subsector then subsumed even smaller operational units, such as "battalions," "brigades," or "groups," according to the Travelstar documents.

Between May and October 2006, the sectors of Anbar province were listed as ar-Rutbah, al-Gharbiyah (or "Western"), al-Awsat (or "Central"), at-Ta'mim, ar-Ramadi, and Fallujah and belts.[6] These are roughly similar to the official GoI geographic districts of Anbar. Our assessment of how these bureaucratic divisions match to a map of Anbar province is displayed in Figure A.2 in Appendix A.

---

[5] Jonathan D. Farley, "Breaking Al Qaeda Cells: A Mathematical Analysis of Counterterrorism Operations (A Guide for Risk Assessment and Decision Making)," *Studies in Conflict and Terrorism*, Vol. 26, 2003, pp. 399–411; Shapiro, 2008; Fishman, 2008.

[6] AQI documents consistently refer to areas surrounding Fallujah as the city's "belts." The same terminology is used in AQI documents about Baghdad. It is unclear whether the belts around Fallujah noted in these documents are to any extent synonymous with Baghdad's belts.

The Bureaucracy of AQI in Anbar

AQI structured its bureaucracy similarly at the provincial and sector levels. Both the provincial and western Anbar units had administrative emirs who represented and accounted for each sector's administrative system, and they both had bureaucratic sections (with a specific division of labor), namely: "movement and maintenance," "legal," "military," "security," "medical," "spoils," and "media."

There were differences between the units, however. The western sector bureaucracy had a slightly more elaborate division of labor than the provincial administration, and the provincial group funded a number of ancillary units unseen at the sector level. Specifically, the provincial bureaucracy includeed a Mujahidin Shura Council, ostensibly the council for cooperation between many Sunni insurgent groups, although from the documents it is unclear what this council did. In addition, the provincial bureaucracy contained an Administrative Council, which oversaw all functional bureaucratic departments and may have maintained unique revenue streams; a General Treasury, which appeared to be a shared fund for other AQI provinces or for central AQI leadership; and a Mail section, which we believe controlled a system of couriers. The western Anbar sector also had functional departments not seen in the provincial administration, such as "boats," "relief and depots," and the "Soldiers Chamber (salaries)."

The bureaucratic structure AQI employed is called a multidivisional or "M-form" hierarchy, which creates autonomous geographic units (in our case, the sectors) in the periphery, each of which encapsulates a wide range of functions.[7] The central units in M-form hierarchies formulate broad policy and strategy and implement their decisions through administrative apparatuses responsible for governing the state. In this case, AQI had a wide array of functional units contained in both the center (the province) and peripheral (sector) units. Further, a "national" level of AQI bureaucracy may also have existed during this period, given the system we observed in Anbar province. M-form hierarchies minimize contact between the peripheral units, forcing

[7] Oliver Williamson, *Markets and Hierarchies: Analysis and Antitrust Implications,* New York: Free Press, 1975.

interperipheral relations to be mediated by the center. Perhaps not coincidentally, the CPA and MNF-I also were organized in M-form hierarchies, with different divisions controlling the provinces of Iraq.[8]

A Change in Emirs

Over time, both the individual appointed to be administrative emir of AQI's provincial administration and the administrative emir of the western sector changed in response to events, but this change in personnel produced only slight changes in accounting procedures in either case. This may indicate strictly defined and enforced accounting rules in the organization. Regarding the provincial administrative emirs, one named Firas held the office from 2005 to May 2006, but from May 2006 forward, an individual named ʻImad held the office. There were no apparent gaps in the accounting as a result, which may indicate that the previous emir either moved or was transferred to another part of the organization.

Division of Labor

The Administrative Council of AQI in Anbar oversaw all functional departments, which include security, legal, medical, mail, media, prisoners, support battalion, and military—each of which was associated with a named individual in the documents. Similarly, each sector had its own set of functional departments, each associated with a specific individual emir.

In addition to the provincial administration's division of labor, the western sector divided its labor according to three structural levels: (1) the sector level, (2) the "brigade" level, and (3) the "group" level. Specifically, each sector had a general emir and separate emirs for each functional department, including the bureaucratic departments listed above (see Figure 3.1). Each person responsible for the administrative duties in a sector had the authority to oversee its subsidiary brigades, but it is unclear whether this person was the sector general emir, the sector administrative emir, or another person; the Arabic refers only to the masʻul [person responsible] for administration. This person

---

[8] Alexander Cooley, *Logics of Hierarchy: The Organization of Empires, States and Military Occupations*, Ithaca, N.Y.: Cornell University Press, 2005.

also appointed the brigade administrator. These administrators were responsible for passing financial and military reporting up to higher levels of organization, creating a vast paper trail of AQI's activities. The evidence for this division of labor at the sector level has also appeared in documents pertaining to the Tarmiyah area north of Baghdad, and thus we believe it is likely that this structure applied to the other AQI provincial organizations outside Anbar.

Financial Flows

Figure 3.1 shows a composite picture of AQI's financial flows in Anbar province. The Anbar administrative emir obtained revenues from three sources:

- each sector's administrative emir (who obtained revenues from the AQI groups in his area) and the border administrative emir
- the general emir, who obtained funds from the General Treasury
- the Spoils Groups.

With these revenues, the administrative emir funded the expenses of the general emir, a portion of the six sectors' administrative emirs (who pay the expenses of their sectors), the border administrative emir, and the Anbar Administrative Council. With funds received from the administrative emir, the general emir paid the expenses of the General Treasury and the Mujahidin Shura Council. The general emir also allocated small amounts of funding to two southern sectors, one of which is indicated as being in Basra.

Revenue-Generating Activities

The provincial revenues were tallied on two different master sheets in the document collection. The first sheet corresponds to Firas's administrative period (June 2005 to May 2006), and the second corresponds to 'Imad's administration (May 2006 through September 2006). We analyzed these two periods separately because of a difference in the administrators' accounting; the first broke out individual revenue transactions by their detailed description, whereas the second used only the

Figure 3.1
AQI Financial Flows in Anbar Province

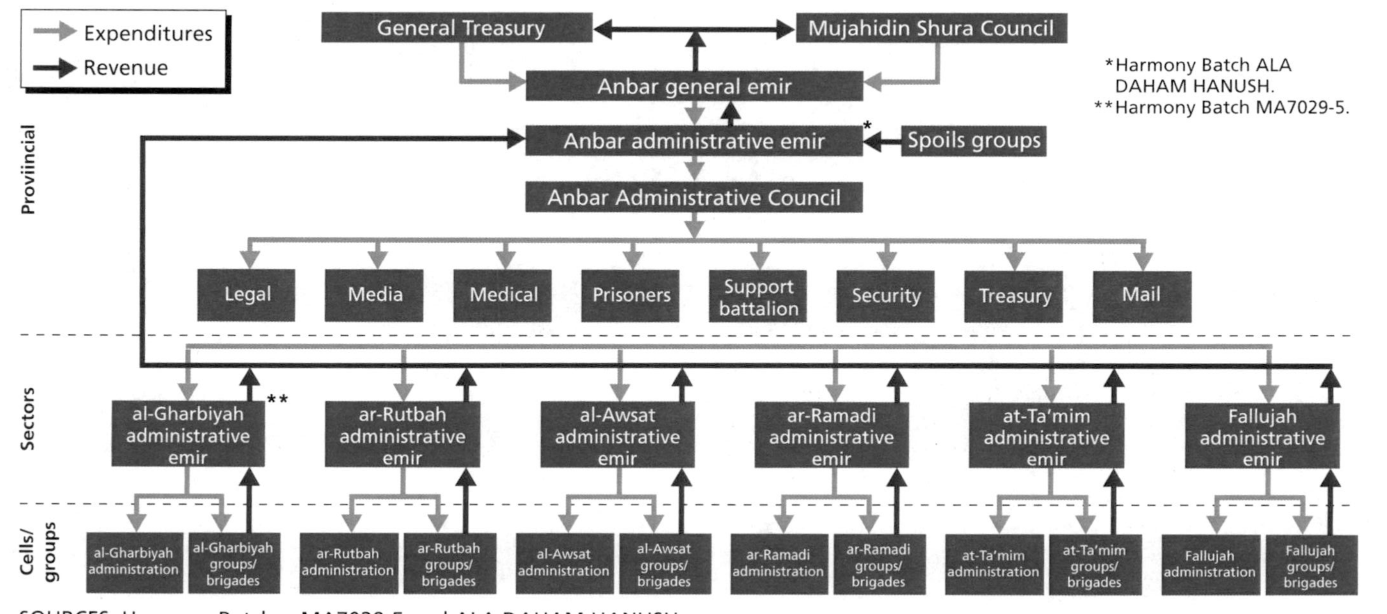

SOURCES: Harmony Batches MA7029-5 and ALA DAHAM HANUSH.

NOTES: Review of the information in the provincial administration's spreadsheets leads us to believe that the other five sectors besides al-Gharbiyah are structured in the same manner as al-Gharbiyah. The link between the spoils groups and the Administrative Council is unclear, but the spoils groups provide the council with substantial funding.

RAND *MG1026-3.1*

broad categories "spoils and sales," "donations," and "transfers from the sectors."

From June 2005 to May 2006, AQI's Anbar administration raised nearly $4.5 million or roughly $373,000 per month (Figure 3.2). AQI in Anbar was financially self-sufficient. This is consistent with the concept of al-Qa'ida's franchising strategy as described by Kiser (2005), who argued that al-Qa'ida central provides seed capital to its franchises but pushes them to quickly become financially self-sufficient.[9]

The group obtained more than 50 percent of its revenue from selling what appear to be stolen goods, most of which were highly valuable capital items, such as construction equipment, generators, and electrical cables.[10] In contrast to what has been suggested in much news reporting, the data do not support the claim that the group was largely financed by selling stolen oil, as the revenue garnered from oil appears to be fairly negligible in the context of total group revenues at this level of administration (note that oil revenues are not shown separately in Figure 3.2 but are instead a small portion of the stolen goods entry in the figure).[11] However, it is also entirely possible that oil revenues were garnered by one of the many Anbari AQI sectors that we do not have data on, as well as by the national AQI administration, which had a number of purported ministries and claimed to have a specific "oil minister."[12]

---

[9] Steve Kiser, *Financing Terror: An Analysis and Simulation to Affect Al Qaeda's Financial Infrastructures,* Santa Monica, Calif.: RAND Corporation, RGSD-185, 2005.

[10] We assume that these goods were all stolen, as in some cases they are explicitly denoted as "booty" and in others they simply appear on the group's balance sheets without any record of their being paid for. It is therefore also possible that goods that simply appear on the books were donated in Iraq or donated abroad and imported as part of a trade-based money-laundering scheme.

[11] Oppel, 2008; Lennox Samuels, "Al Qaeda Nostra," Newsweek Web Exclusive, May 21, 2008.

[12] "Al-Furqan Media Wing Declares the Members of the Cabinet of the Islamic State of Iraq," April 19, 2007, as cited in Kohlmann, 2007; "Oil Smuggler and Al Qaida Supplier Arrested in Bayji," *Al Mashriq Newspaper,* November 22, 2007, as cited in Phil Williams, *Criminals, Militias, and Insurgents: Organized Crime in Iraq,* Carlisle, Pa.: Strategic Studies Institute, U.S. Army War College, June 2009.

Figure 3.2
Anbar Province Revenues, June 2005–May 2006

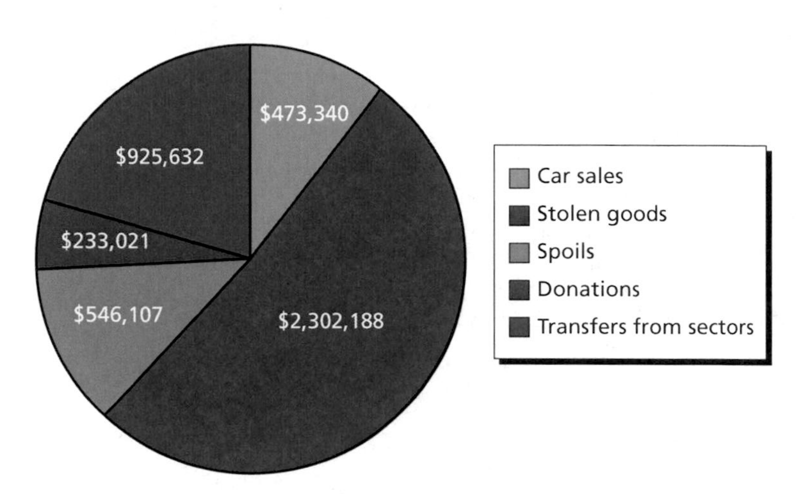

SOURCE: Harmony Batch ALA DAHAM HANUSH.
RAND *MG1026-3.2*

AQI's provincial administration also collected a substantial amount of revenue that was raised by the sectors, constituting 21 percent of the provincial administration's total revenues. This indicates that each sector had its own revenue-generating activities, and it appears that the sectors with excess cash on hand were asked to pass funds up to the provincial administrator. Although constant reporting allowed this to be possible, it is likely that the sectors evaded such revenue sharing, as it would be difficult for the province to incentivize them to fully cooperate in this manner.

Car sales and spoils each constituted just over 10 percent of provincial revenues. Spoils are the proceeds from goods taken from those that AQI designated as "apostates" by the group's Islamic law section. We believe that these individuals were either Shi'a or collaborators with the local government or U.S. forces. Finally, explicit donations accounted for only 5 percent of total revenues.

From June to November 2006, the provincial administrative unit dramatically increased its revenues, making $4.3 million over the period or $860,000 per month (Figure 3.3). The group realized this dramatic uptick in revenue—from $373,000 a month to $860,000—

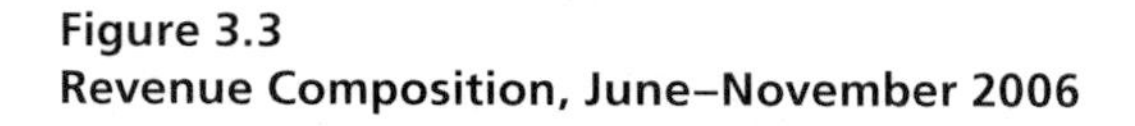

Figure 3.3
Revenue Composition, June–November 2006

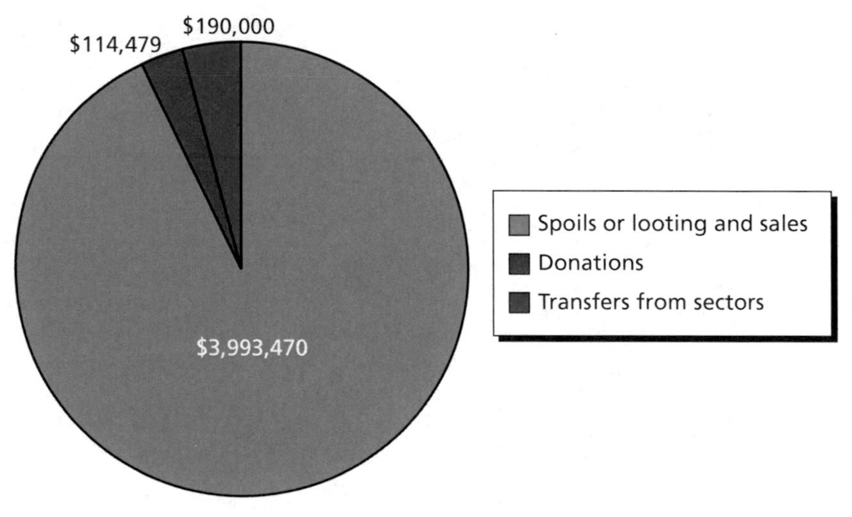

SOURCE: Harmony Batch ALA DAHAM HANUSH.
RAND MG1026-3.3

through an increase in all of the observed revenue streams except transfers from sectors, which declined substantially; part of this decrease may have been due to the increasing centralization of AQI in Anbar at that time. Revenue from spoils and from sales (accounting for both cars and stolen goods) increased in a disproportionate manner. Spoils and sales increased from 74 percent of revenues in the earlier period to 93 percent in the later period, whereas transfers from sectors decreased as a share of total revenues. The rapid shift in revenue sources may indicate that AQI became more aggressive in ordering AQI fighters to collect spoils and to leverage Anbar's commercial economy for its own gains, specifically to increase its offensive capabilities. This apparent offensive shift in AQI's revenue-seeking behavior may have exacerbated the already acute concerns of tribal sheiks involved in smuggling and highway robbery, such as Sheik Sattar Abu Risha, and likely contributed to the coalescence of the Awakening in the fall of 2006.

The provincial administration collected revenue mostly from the sale of stolen commercial goods and did not collect much revenue from black market fuel sales, large-scale extortion, or direct taxation of the populace. The financing of the western sector was similar, where AQI funded itself through the sale of cars and lesser-valued commercial goods. The western group focused on taking spoils and booty from those whom the group deemed to be apostates, following their ideology in strict, exploitative terms. The AQI financing model, therefore, could be characterized as local, religiously radical, and politically destabilizing; it encompassed vigorous independence (specifically, local financing), exploitation based on radical religious belief, sensitivity to the perception of the average Sunni, and an explicit attempt to subvert the power of a number of key tribes.

Another noteworthy aspect of AQI's revenue activity was the demarcation of revenues being collected by "spoils groups." This finding warrants deeper examination, because this notation may indicate that AQI had specifically designated units for extortion and theft. As discussed above, spoils appeared to be goods and items taken from apostates in ways that are ruled as legitimate by AQI legal emirs. However, this activity was clearly extortion or theft undertaken by either AQI members or criminal elements that were co-opted by AQI.

Revenues were commonly generated by units that were directly administered by the sectors, although we observe that large revenue streams seemed to pass directly to the Anbar administration. This could mean that the Anbar administrator had business assets that operated within the sectors or that sector assets were passed up to the province administration by rule. Although we do not know which of these occurred, it is likely that sector-based groups passed large revenue streams to the provincial administration because the higher administrative unit has bureaucratic precedence.

Expenditures and Funding of Subsidiary Units

AQI's Anbar administrator funded a multitude of affiliated units; in particular, he sent over 50 percent of his funds to the six subsidiary sector-level administrators (Figure 3.4). The provincial administrative section used 11 percent of total spending to fund administrative operations. In comparison, direct military activities—

Figure 3.4
Composition of Expenditures by AQI in Anbar

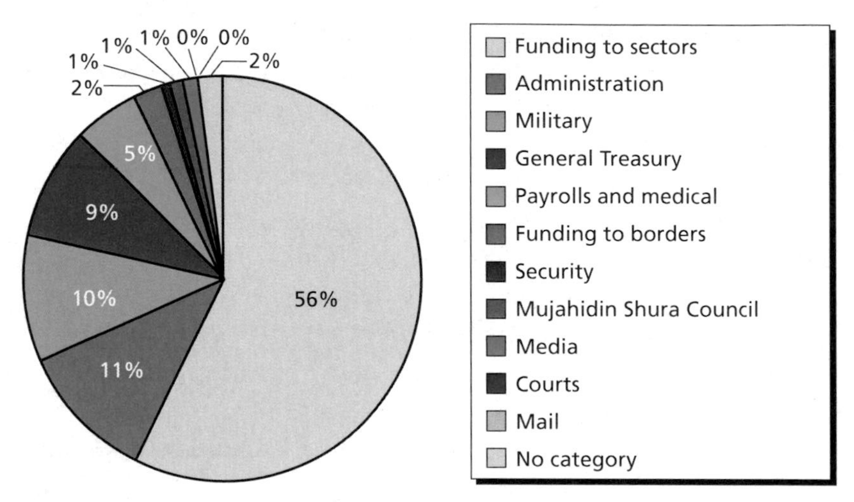

SOURCE: Harmony Batch ALA DAHAM HANUSH.
NOTE: Percentages do not sum to 100 percent because of rounding.
RAND *MG1026-3.4*

constituting an attached military battalion and a military procurement section—received only 10 percent of the group's spending. The General Treasury received 9 percent of total revenue. Interestingly, payrolls and medical constituted only 5 percent of spending for the province, while the other administrative subsections, such as courts and mail, received less than 1 percent of the total budget.

The sectors that are farthest from the western border and simultaneously most central to AQI's operations, Fallujah and ar-Ramadi, received the most in net money transfers (Figure 3.5). This indicates, first, that the organization prioritized operations in these areas, but, second, that the group generated most of its revenues away from the cities where it focused its operations. This could be explained by illicit trade and large-scale theft being most viable or profitable in rural areas along smuggling routes or by the group's inferred preference to keep its business operations away from the scrutiny of most of the Anbari public.

Figure 3.5
Flows to and from Subsidiary Sectors

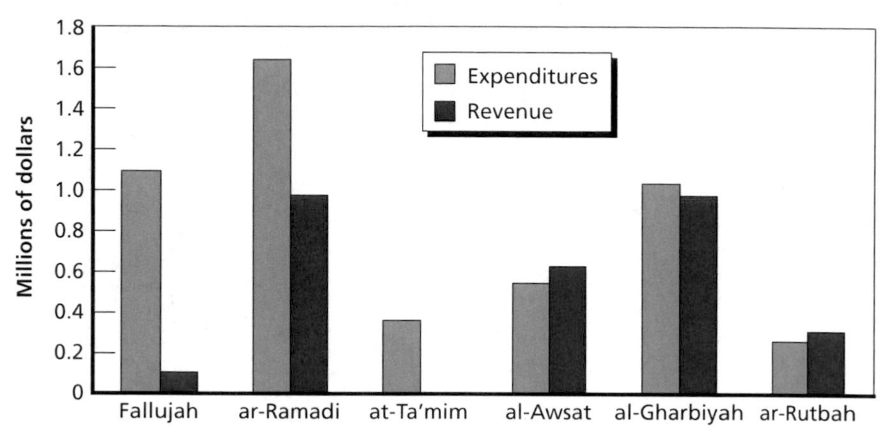

SOURCE: Harmony Batch ALA DAHAM HANUSH.
NOTES: The figure includes only expenses and revenues for months in which all six sectors were accounted for. The figure does not account for revenues and expenses that were tallied but not ascribed to a specific geographic area or individual identified with an area.
RAND *MG1026-3.5*

The provincial administration functioned as both a unit of financial oversight and financial facilitation, enabling high levels of operational tempo across large geographic areas through the sharing of revenue streams. It did so by allocating funding across sectors, from those that raised more revenue relative to their expenses to those that raised less, and managing large revenue streams that otherwise might not have been contributed to a shared pool.

Analyzing the funding flows to the sectors, we find that the monthly transfers to the different sectors largely rose and fell in unison, again with the exception of ar-Ramadi and Fallujah (see Figure 3.6). Also, it is apparent that as revenues increased after May 2006, nonsector spending increased disproportionately. Such an increase may indicate that the administrative emir preferred to send excess funding to units that he directly controlled, which may have increased his power relative to the sectors or relative to other province administrators. This increase

Figure 3.6
Monthly Transfers to Sectors and Nonsector Spending

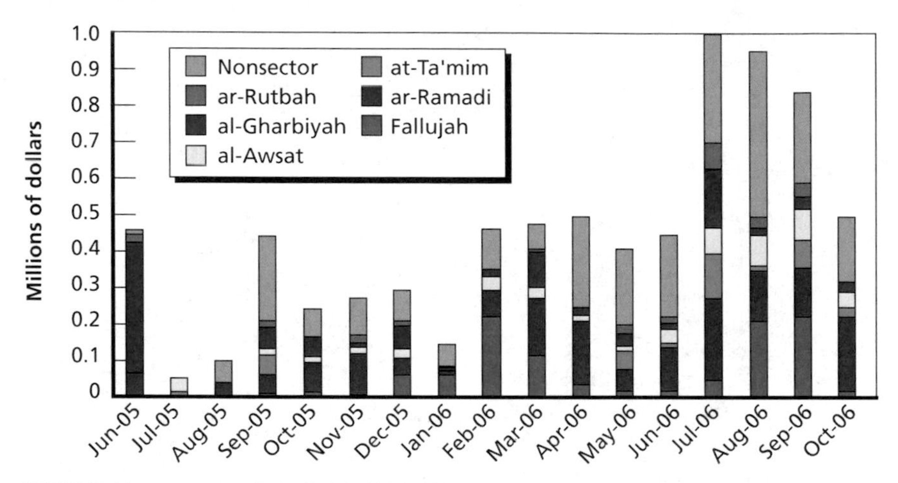

SOURCE: Harmony Batch ALA DAHAM HANUSH.
NOTES: This chart excludes the lump-sum transfers to the General Treasury in March, July, and August 2006, as we believe these funds were sent out of the province. The large, anomalous expense in al-Gharbiyah in June 2005 was to purchase safe houses.
RAND *MG1026-3.6*

appears in the Anbar-level funding to military, administration, and payrolls as opposed to financial flows to sectors for the sectors to distribute.

The administrator exhibited a strong and consistent preference to move money rather than hold it. He sent extremely large amounts on the same day he received them, usually to the provincial administrative units, the local military wing, or the General Treasury. The Anbar administrator kept only between $25,000 and $250,000 in cash on hand throughout the period, and he transferred large excess revenues to the General Treasury. In light of this behavior, it is possible that the General Treasury was a shared fund for AQI's senior leaders or al-Qa'ida outside Iraq.[13] The Anbar emir transferred a net sum of $390,000 to the General Treasury over the entire period, which leads us to believe that Anbar was a net contributor to AQI nationally

[13] The documents we examined give no further clues as to the purpose of the General Treasury. However, more information may exist in the document set.

or perhaps to al-Qa'ida central during this period. The fact that the Anbar administrative emir transferred funds to such far-flung places as Mosul, the border sections, and Basrah bolsters the argument that Anbar was a highly profitable locale for AQI.

Administrative Decisionmaking

AQI revenues and expenditures in Anbar province exhibited a strongly positive correlation at the monthly level, with a correlation coefficient of 0.97 (Figure 3.7).[14] The administrator of Anbar balanced his books and updated his master accounting sheet biweekly. As noted above, the administrator saved very little revenue. We believe that this behavior

Figure 3.7
AQI Monthly Expenditures and Revenues, Anbar Province

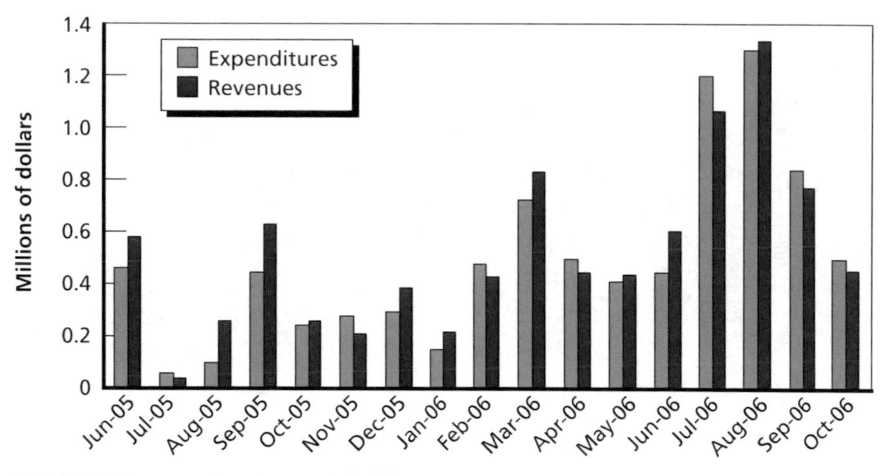

SOURCES: Harmony Batch ALA DAHAM HANUSH.
NOTES: This chart includes the large transfers to the General Treasury that occurred in March, July, and August 2006.
RAND *MG1026-3.7*

[14] Correlation coefficients range from –1.0 to 1.0, with –1.0 occurring when variables are perfectly negatively correlated and 1.0 occurring when variables are perfectly positively correlated.

was intended to provide a high level of operational tempo across the province in the short run, because it was a known goal of AQI to sustain high attack levels to sap the Coalition's willingness to remain in Iraq. At the same time, the administrative emir had a distinct function apart from his direct supervisor, the general emir. The administrative emir was a facilitator for the needs of the sectors and his subsidiary offices within the provincial administration, as only occasionally did the general emir explicitly order the disbursement of funds.

The data allow us to observe AQI at its zenith in Anbar in mid-2006. The administrative emir added the sectors of ar-Rutbah, at-Ta'mim, and Fallujah in May 2006 when the group's revenue streams were markedly increasing, adding to the extant sectors of ar-Ramadi, al-Awsat, and al-Gharbiyah. The documents also show the reaction of the administration to the adversity it faced when the Awakening organized and attacked AQI in Ramadi in late September and early October 2006. During this period, the provincial administration provided increased funding to the ar-Ramadi sector, where CF were also conducting intensive military operations against AQI. Furthermore, on October 16, 2006, the administrator eliminated the at-Ta'mim, Fallujah, and ar-Rutbah sectors from his ledgers and renumbered those that remained. The three eliminated sectors provided less than 5 percent of the provincial administration's revenues over the period. Accordingly, we believe that this action indicates a choice of the administrative leaders to consolidate the areas being funded by AQI higher leaders because of the intense fighting taking place against Awakening forces. This type of strategic decisionmaking by AQI's administration would be possible only through a managed, hierarchical organization. However, it is also the case that the ISI was established by AQI leaders on October 15, 2006, and this may have been accompanied by administrative changes mandated by ISI senior leaders. Whether the change in the sectors was due to fighting SAA or because of new rules in accordance with the creation of ISI cannot be discerned from the available data.

CHAPTER FOUR

The Economics of AQI's Compensation

AQI's financial ledgers provide a unique opportunity to analyze the economics of a contemporary militant group, focusing on individual compensation and on the relationship between the group's expenditures and attacks. In this chapter, we measure the relative economic benefits of being an AQI member by comparing AQI members' salaries to reported income levels of Anbari households from available survey data. We then compare the risk of death that prospective AQI members faced with the risk of death for members of the broader Anbari population to explore whether AQI's compensation reflected this risk in a purely economic sense. In the next chapter, we statistically test whether changes in expenditures were associated with changes in attack levels. These analyses provide a deeper understanding of the decisionmaking of militant groups and the incentives facing their constituent members.

Payrolls

According to the captured documents, AQI paid its single members throughout Anbar 60,000 Iraqi dinars per month in 2006 (roughly $41 per month, or $491 per year in nominal terms). In addition, members each received 30,000 dinars per month for each of their dependents, including parents, unmarried sisters, brothers under the age of 15, unmarried daughters, and sons under the age of 15 (roughly $20 per month, or $245 per year).[1] A handful of members were listed as

[1] We converted dinars to dollars at a rate of 1,469 in 2005 and 1,467 in 2006, the average monthly rates for each year (Central Bank of Iraq, "Key Financial Indicators for February 17, 2010," Excel spreadsheet, Baghdad, Iraq, 2010).

holding other jobs. In these cases, the administrator subtracted the salary they received from their other employers from the assistance they received from AQI. The compensation structure was flat; pay was not differentiated by type or skill of labor, or position in the group. The group maintained a model that provided a basic amount of economic security for members based on a rough metric of household need—family size—and nothing more.

One way to compare the relative compensation of AQI members and other Iraqi civilians is to consider Iraq's per capita gross domestic product (GDP). Iraq's per capita GDP in 2006 was estimated to be $1,770.[2] In comparison, members with no dependents were receiving $491 per year during the period covered by these documents. According to these figures, salaries for AQI personnel in Anbar were equivalent to ony 28 percent of average, nationwide per capita GDP.

However, the Iraqi GDP includes oil revenues, which may not trickle down into household income. Therefore, an alternative, and better, measure is to compare AQI Anbar salaries with survey data on income in Anbar (Table 4.1). We put all figures in real terms because even though the time period is short, Iraq went through a period of very high inflation from mid-2004 through mid-2007. Between July 2004 and June 2007, the monthly consumer price index (CPI) increased by almost 204 percent, or about 46 percent per year.[3]

We compare AQI salaries to Anbar salaries in 2005 and 2007. Although the AQI ledgers date from 2006, available internal evidence suggests these salary levels were the same in 2005. Although we do not have survey data for overall Anbari compensation for 2006, we do have survey data for 2005 and 2007.

AQI compensated its members poorly compared to the compensation that the general population earned. According to both 2005 and 2007 data, mean per capita AQI compensation was less than half that

---

[2] International Monetary Fund, "Iraq: Third and Fourth Reviews Under the Stand-By Arrangement, Financing Assurances Review, and Requests for Extension of the Arrangement and for Waiver of Nonobservance of a Performance Criterion," Washington, D.C., February 23, 2007.

[3] Central Bank of Iraq, 2010.

Table 4.1
Monthly AQI Compensation Versus Monthly Anbar Compensation

	Anbar					AQI
	2004	2005	2007	2008 Urban	2008 Rural	2006
Per capita household income (Iraqi dinars)						
25th percentile	43,829					34,286
Median	76,164					37,500
Mean		93,415	91,260	99,117	94,882	42,876
75th percentile	130,424					60,000
Household income (Iraqi dinars)						
24.6th percentile			382,162			
25th percentile	327,908					60,000
47.7th percentile			535,027			
Median	569,280					150,000
Mean		663,246	646,006	683,913	645,201	162,766
73.3rd percentile			764,324			
75th percentile	959,929					240,000
Per capita household income (U.S. $)						
25th percentile	30.16					23.37
Median	52.42					25.56
Mean		63.59	72.72	83.08	79.53	29.23
75th percentile						40.90
Household income (U.S. $)						
24.6th percentile			304.51			
25th percentile	225.68					40.90
47.7th percentile			426.32			
Median	391.80					102.25

Table 4.1—Continued

	Anbar					AQI
	2004	2005	2007	2008 Urban	2008 Rural	2006
Mean		451.49	514.75	573.27	540.82	110.95
73.3rd percentile			609.02			
75th percentile	660.65					163.60

SOURCES: Ministry of Planning and Development Cooperation, 2005, for 2004 data; United Nations World Food Programme, Iraq Country Office, and COSIT, Ministry of Planning and Development Cooperation, Iraq, *Food Security and Vulnerability,* United Nations World Food Programme, 2006, for 2005 data; COSIT, Ministry of Planning and Development Cooperation, Iraq, KRSO, Nutrition Research Institute, Ministry of Health, Iraq, and United Nations World Food Programme, Iraq Country Office, 2008b, for 2007 data; Crane et al., 2009, for 2008 data; Harmony Batch MA7029-5 for AQI data; and Central Bank of Iraq, 2010, for exchange rate and inflation data.

NOTES: All figures are real in terms of 2006 Iraqi dinars. We used the annual average of monthly Iraqi CPI data to convert figures to real terms, and then the annual average of monthly exchange rate data to convert the figures to dollars. 2005 figures are in terms of per capita income rather than per capita household income. We calculated 2008 per capita figures by dividing household income by household size.

received by the general population of Anbar. Mean household compensation was only about one-quarter of civilian compensation, presumably because AQI members had larger households than average. Although AQI's pay structure could have encouraged people with large households to join, the low level of pay strongly suggests that members were not motivated purely by financial gain.

Risk

A simple comparison of salaries may be misleading, as AQI members were at higher risk of death or imprisonment than their nonmilitant counterparts. In this section, we assess the level of risk by comparing the relative risk of death among AQI members and that of the general public in Anbar province. We then compare expected lifetime earnings to see how much worse AQI members fared in terms of compensation because of the added risk that came with membership in the group. We recognize that it is unlikely that a potential AQI recruit knew the risk of death in advance. However, it is likely that such a person would know that by taking up arms, he would be increasing his risk of death substantially. One value of our analysis is that we show just how substantial this increase was.

Anecdotal evidence suggests that risk of death or capture is generally much higher among militants, and the data allow us to quantify the risk. We use the data from our documents and from the Iraq Body Count (IBC) project to estimate the different mortality rates of AQI members and the Anbar male population in 2006. Although there are no authoritative counts of Iraqi civilian deaths since the start of Operation Iraqi Freedom, data from the IBC are widely used as one source. Among six sources cited by the Congressional Research Service, the IBC estimates were in line with most others.[4] Because there is no guarantee of perfect accuracy when using the IBC figures, we present several different estimates based on them, as explained in the notes to Table 4.2. We used data from the IBC project because that source is one of the most widely cited and because it contains data disaggregated by geography, time, age, and sex.

Military actions appear to have been effective in making AQI membership a risky endeavor in western Anbar province. Roughly half the individuals on the rosters of soldiers from the western Anbar documents are listed as either captured or killed. Because the group continued to pay support to families of the captured and killed, two-thirds of its personnel budget went to these families and not to active members.[5]

Furthermore, the documents show that in late 2006, between 23 and 30 percent of all living members on the western Anbar payroll list were in detention.[6] The fact that the families of the detained insurgents

[4] Hannah Fischer, *Iraqi Civilian Deaths Estimates,* CRS Report for Congress RS22537, Washington, D.C.: The Library of Congress, August 27, 2008.

[5] Harmony Batch MA7029-5.

[6] Harmony Batch MA7029-5.

Table 4.2
Violent Mortality Risk for AQI Members and Anbar Male Population, 2006

	AQI Members	Anbar Men Age 18–48
Number	251	295,971
Violent deaths		
Estimate 1	87	1,075
Estimate 2		958
Estimate 3		773
Violent deaths per thousand per year		
Estimate 1	173	3.63
Estimate 2		3.24
Estimate 3		2.61

SOURCES: United Nations World Food Programme, Iraq Country Office, and COSIT, 2006; COSIT, KRSO, and the World Bank, 2008; Crane et al., 2009; Iraq Body Count Project, 2010; Harmony Batch MA7029-5 and document MNFA-2007-000562.

NOTES: We estimated the male population age 18–48 for Anbar province in 2006 using data on the proportion of males age 18–48 in 2008 reported in Crane et al., 2009, and data on the Anbar population, household size, and number of men per household in both 2005 and 2007 as found in two Food Security and Vulnerability Analysis surveys (U.N. World Food Programme and COSIT, 2006; and COSIT, KRSO, and The World Bank, 2008). Specifically, we divided population by household size to get the number of households and then multiplied by the number of men per household to get the number of men in Anbar province in 2005 and 2007. We then multiplied this by the proportion of men age 18–48 in 2008 to get an implied number of men age 18–48 in 2005 and 2007. Finally, we took the average of these two figures to get our 2006 population of Anbar men age 18–48. The 2005 figure was 285,010, and the 2007 figure was 306,932.

We estimated the number of violent deaths among men age 18–48 from data provided by the IBC project. The IBC data reported the total number of civilian violent deaths in Anbar in 2006 but included demographic information on only a portion of the people who died violent deaths. We estimated the number of violent deaths among the demographic of men age 18–48 in three ways. The first way was to assume that all deaths, even those known to be women or out of our age group, were in our demographic. This is the most conservative method and gives the highest estimate of deaths per thousand population. This is the figure we use in the text in our discussion of violent death rates, as it also can proxy for men who died violent deaths but who the IBC was unable to record. The second estimate calculated the proportion of male deaths among the IBC deaths for which the gender was known, assuming all males who died were between age 18 and 48, and multiplied this proportion across the entire body count. The third estimate counted the proportion of males known to be age 18–48 in the set of deaths for which any age or gender information was known. This proportion was multiplied by the entire body count for Anbar in 2006. In cases where only an estimated age range was

Table 4.2 Notes—Continued

known (for example, 13–19), we included any individual as being in our demographic as long as some part of the age range overlapped with ours. Estimating the deaths of males age 18–48 gave us a range of estimates that increased the chances that we would not be undercounting such deaths.

Deaths of AQI members are assumed to have taken place between the group's creation in October 2004 and the date of the document in November 2006. All AQI deaths are assumed to have been violent.

receive money from AQI and the detained cannot engage in enemy activity compounds the effect of detention on AQI's ability to operate. The potential benefit of keeping suspected AQI operatives detained as a COIN or counterterrorism measure is high, but this must be tempered by the consideration that detaining the innocent may fuel contempt for Coalition Forces and the Iraqi state.

AQI members suffered an extraordinary rate of death, with 87 out of 251 members on the western Anbar sector personnel rolls listed as dead (Table 4.2). Although the documents are not specific, we assume that this represents deaths since AQI was formed in late October 2004 through the date of the documents in early November 2006. This works out to 173 violent deaths per thousand members per year. In contrast, our upper-bound estimate of the violent death rate of the general Anbar male population is 3.6 violent deaths per thousand inhabitants per year. The comparison shows a staggeringly greater relative risk of violent death for AQI members, who were 48 times more likely to die violently than the average Anbari. As explained further in the notes to Table 4.2, 3.6 deaths per thousand is actually the violent death rate of all civilians in Anbar, but by using this figure and considering all these deaths to be of men age 18–48, we are not only providing the most conservative estimate but also implicitly counting any violent deaths of our target demographic that went unrecorded. We considered insurgent age to be 18–48, although widening or narrowing this range will not change the magnitude of the difference we found.

We caution that we cannot vouch for the accuracy of the AQI figures. However, these tallies of killed, captured, and active members were kept by a higher-level official, the administrative emir, responsible for making sure personnel were paid. The fact that compensation was an important part of organizational cohesion suggests that these documents are at least roughly accurate and usable for analysis.

With these figures, we can further explore the extent to which the additional risk of joining AQI is not adequately compensated by looking at the implicit expected lifetime earnings. We start with the per capita earnings figures and then consider household earnings, comparing AQI earnings to 2007 Anbar earnings.[7] We make all comparisons in terms of real 2006 values. We base this initial analysis on the assumption that an AQI member or a civilian loses his income if he is killed. As discussed above, the documents indicate that this is not the case; the families of AQI members continued to receive compensation when the member was captured or killed. Although the promise of continued compensation may not have been fully credible, and also may not have been honored by AQI, we must examine both assumptions of AQI members receiving their full income regardless of death or a complete loss of income on death. In the next three sections, the assumption that income is lost upon death provides a lower bound for expected lifetime earnings.

Per Capita Household Earnings

As shown in Table 4.1, mean per capita monthly household earnings for average Anbaris were $73 in 2007. This translates to $873 in income per person per year. In contrast, mean per capita household earnings for AQI members in 2006 were $29 monthly, or $351 per year. On the surface, this would suggest that AQI members were not compensated for their risk, earning only 40 percent as much as the average civilian Anbari.

Adding the effect of differential death rates makes the salary comparision even worse for AQI members. As shown in Table 4.2, the civilian risk of violent death was 0.36 percent per year. In contrast, the AQI risk of violent death was 17.3 percent per year. This means that for both cases, it was likely that a person would not earn the average salary because of the risk of violent death. Assuming that (a) these

---

[7] The results remain similar in magnitude if we compare AQI earnings to 2005 Anbar earnings.

annual totals remained constant in real terms and (b) an Anbari or an AQI member would expect a 31-year career at their respective salaries, working from age 18 to age 48, a civilian Anbari could expect to earn dramatically more than an AQI member during a lifetime of work. Adjusting for the risk of violent death, a civilian could expect to earn $25,547 over his lifetime, whereas an AQI member could expect to earn only $1,672 over his lifetime.[8] Put differently, an Anbari civilian could expect 15 times higher lifetime earnings than an AQI member.

Household Earnings

Household earnings tell a slightly different story for civilian Anbaris and AQI members because AQI members had larger families. In our comparison, we considered AQI household earnings to have been earned entirely by the AQI member, but civilian household earnings could have been earned by the head of household and others in the household. We therefore make the strong assumption that civilian household earnings were earned entirely by the head of household, the most conservative assumption for comparing expected lifetime earnings. If we were to assume that some of the household earnings came from others in the household, this would make expected civilian earnings compare even more favorably to AQI earnings than in the comparison we discuss below.

Based on data in Table 4.1, mean Anbari civilian household earnings amounted to $6,176 annually, whereas mean AQI household earnings amounted to $1,331 annually. As before, adjusting for the risk of death, keeping these values in real terms, and assuming a 31-year working life, Anbari civilian households would earn $180,843 over their lifetimes, whereas AQI households would earn just $6,347 over their lifetimes. Put differently, Anbari civilians could expect more than 28 times higher household earnings over their lifetimes.

[8] The specific formula is $\sum_{t=1}^{30}(1-r)^t S_A$, where t is the year of work, from 1 to 31, r is the annual risk of death (0.0036 for civilians and 0.173 for AQI fighters), and S_A is the annual salary in real terms for either Anbari civilians or AQI fighters. The annual salary is set to the base year salary as shown in Table 4.1, and the calculation assumes that the annual salary is collected at the end of each year.

Considering Discount Rates and Lifetime Payments

An additional factor that could tilt the calculations is how much AQI members and other Anbaris valued income in the present over income in the future. Such a measure is captured in a discount rate. For example, if a person is promised $100 today and the same amount one year from now, in real terms, that person might still place a higher value on the money today. If he considered $110 next year to be equivalent to $100 today, that would compute to a discount rate of 10 percent.[9]

Empirical analyses of human behavior have revealed a wide range of individual discount rates.[10] One study of U.S. military personnel found nominal rates of 10.4 percent to 21 percent for officers and 35.4 percent to 57.2 percent for enlisted personnel.[11] The value of that study was that, unlike most other studies, it showed the results of people making real choices over when they receive large amounts of money. Most other research on individual discount rates has involved controlled experiments.

In fact, there is no discount rate that could equalize AQI and Anbari incomes, keeping the assumption that the income of AQI members is cut off upon their death. This is because AQI salaries were lower to start with, and even lower in expected lifetime terms after factoring in the risk of death. An extraordinarily high discount rate

---

[9] Formally, 110/(1.1) = 100.

[10] Individual discount rates are different from social discount rates, which are the rates at which society as a whole values (or should value) the present over the future. Social discount rates typically are, and should be, lower than individual rates. They are also lower than market interest rates, one measure of the value of the present over the future (Andrew Caplin and John Leahy, "The Social Discount Rate," *The Journal of Political Economy,* Vol. 112, No. 6, December 2004, pp. 1257–1268). One attempt to compute social discount rates for a broad range of countries found a mean of 6.8 percent with a standard deviation of 3.9 percent (Joice Valentim and Jose Mauricio Prado, "Social Discount Rates," Working Paper, University of São Paulo, Brazil, and IMT Lucca Institute for Advanced Studies, Italy, May 6, 2008). Other work has found much lower rates for investments with payoffs in the range of 100 to 300 years (Richard G. Newell and William A. Pizer, "Discounting the Distant Future: How Much Do Uncertain Rates Increase Valuations?" *Journal of Environmental Economics and Management,* Vol. 46, 2003, pp. 52–71).

[11] John T. Warner and Saul Pleeter, "The Personal Discount Rate: Evidence from Military Downsizing Programs," *The American Economic Review,* Vol. 91, No. 1, March 2001, pp. 33–53.

of 1000 percent would still leave per capita household expected Anbar lifetime incomes more than three times higher than per capita household expected AQI lifetime incomes.[12] However, these computations actually underestimate the lifetime incomes for AQI members.

The records indicate that AQI paid the dependents of members even after those members were killed. Members were promised that their dependents would continue to receive the total amount that they had been receiving before death. This served as a form of insurance, and, to our knowledge, no such equivalent form of insurance existed for the average Anbari. Even if a member were killed, his household income would stay the same. This means that from a household's financial perspective, it was as if the member had no chance of being killed—the family would continue to receive compensation as if the member were still alive. Even with an implicit death rate of zero, AQI salaries were unambiguously worse than average Anbari earnings, even after discounting. All available evidence indicates that AQI members were poorly compensated relative to average Anbar residents, suggesting that financial rewards were not a primary motivation of membership.

We conclude this section with one final computation that the salary data allows us to make. The difference in expected lifetime earnings can be used to measure the value of the non-financial motivations of the average AQI member. Because we do not know the applicable discount rate, we include only death rates in our computation. Given a risk of death of 0.36 percent per year, an average Anbari could expect lifetime earnings of $25,547 per household member. Recall that AQI salaries continued after death, equivalent to no risk of death. On this basis, an average AQI member could expect lifetime earnings of $10,873 per household member. Thus, the non-financial motives were worth giving up $14,675 to an average AQI member, or 57 percent of expected civilian lifetime earnings.

[12] The specific formula is $\sum_{t=1}^{30} \frac{(1-r)^t S_A}{(1+d)^{t-1}}$, where the formula is the same as before but with the addition of d for the discount rate. The discount generally falls between 0 and 1 in ratio terms (between 0 and 100 in percentage terms), but in theory could be any number.

The value of non-financial motives is even greater when calculated on the basis of total household earnings. Given a risk of death of 0.36 percent per year, an average Anbari could expect lifetime household earnings of $180,843. Given an effective risk of death of 0 percent, an average AQI member could expect lifetime household earnings of $41,274. Thus, the non-financial motives were worth giving up $139,569 in household earnings to an average AQI member, or 77 percent of expected civilian lifetime household earnings.

The Many Motivations of AQI Members

People have many possible reasons for joining a militant group such as AQI, including ideological, religious, political, or nationalistic beliefs; tribal issues; the hope of financial rewards; matters of personal honor or revenge; or the simple desire for notoriety. Our calculations, rough as they are, indicate that if AQI members were rational when it came to their finances, financial rewards were not a primary motivation, and that AQI members were not adequately compensated for the additional risk they took on. In fact, we estimate that these other motivations in aggregate were worth forgoing at least 57 percent to 77 percent of expected future income for the average AQI member.

There are two ways by which our finding could be reversed. First, the rewards of AQI membership could have been higher than those reflected in the salary documents. AQI members were known to engage in various types of criminal activity. Although they may have given the proceeds from these activities to the organization, they very well also might have kept some portion for themselves. Second, we have no data on the skills or level of education of AQI members. It could be the case that they were largely unskilled and that the AQI salary was higher than what they could have earned in alternative occupations, even though it was below the provincial average. Given the extraordinary differential and the fact that even lower-income Anbar residents earned more than AQI members (as shown in Table 4.1), however, we think the data provide strong evidence to suggest that AQI members were not compensated for risk, regardless of their skill and educational levels.

CHAPTER FIVE

The Flow of Expenditures and the Pace of Attacks

Many insurgency theorists, military operators, and intelligence officials have posited that the financing of insurgent groups is pivotal for sustaining their operations, and thus their financial systems should be key targets in operations by counterinsurgents. The wide acceptance of this principle led the U.S. National Security Council in 2005 to create the Iraq Threat Finance Cell, a joint Department of Defense and Department of the Treasury group whose primary mission was to increase the quality and availability of intelligence information on insurgency-related financial issues.

Furthermore, there is a wide-ranging acceptance across the U.S. government that using intelligence to follow the finances of terrorists, drug traffickers, and weapons proliferators is a useful way to track these groups and may expose new ways to degrade their capabilities. In policy circles, this type of intelligence and the concomitant policy measures are referred to as "threat finance intelligence" and "counter threat finance," respectively. The analysis presented in this monograph contributes to the threat finance effort by applying statistical techniques that isolate the association between spending and attacks from ongoing trends and conditions in each Anbar sector.

The data in this document collection allow us to test the hypothesis that the pace of AQI's financing is related to the pace of attacks as measured in the U.S. Department of Defense's Significant Activities III (SIGACTS) database. SIGACTS data include all attacks, including those against CF, ISF, and civilians. However, we believe that attacks against the ISF and civilians are undercounted in these data because they were collected by military operators in the field who did not receive many reports about violence against civilians and Iraqi forces. Despite that, SIGACTS is a standard U.S. government database for assessing violence in Iraq and can fairly represent trends and patterns; it is used in the U.S. Department of Defense's quarterly reports, *Measuring Stability and Security in Iraq,* and in publications by the CTC at West Point, and it is transmitted by the Office of the Secretary of Defense to the Special Inspector General for Iraq Reconstruction (SIGIR) for use in SIGIR reports and audits.[1]

We count all attacks in the database as having been perpetrated by AQI. Although SIGACTS does not differentiate the attacks by the group that perpetrated them, AQI was by far the dominant militant group in Anbar during the period of our data, as noted in Chapter Two. Note that we are testing the hypothesis within the narrow scope of AQI's operations in Anbar province at the height of the Iraqi insurgency.

The specific spending variable we test is transfers from the administrative emir of Anbar to the sector level. The money used for these transfers came from a variety of activities, including blackmail, kidnapping, carjacking, and other AQI activities. The specific relationship we test is that between those transfers and total attacks in each sector. In exploring the association of events with the expenditures data, we exclude transfers from the Anbar administrative emir to the General Treasury because we believe that these funds were not immediately used to support AQI in Anbar. We also exclude spending not directed to a specific sector of Anbar. In addition, we exclude spending of money raised within each sector and thus not transferred to the sectors from the provincial administrative emir because we have data on internal

---

[1] For sources that use SIGACTS, see U.S. Department of Defense, *Measuring Stability and Security in Iraq: September 2007 Report to Congress in Accordance with the Department of Defense Appropriations Act 2007 (Section 9010, Public Law 109-289),* Washington, D.C., September 14, 2007; U.S. Department of Defense, *Measuring Stability and Security in Iraq: September 2009 Report to Congress in Accordance with the Department of Defense Supplemental Appropriations Act 2008 (Section 9204, Public Law 110-252),* Washington, D.C., November 4, 2009; and Joseph Felter and Brian Fishman, *Iranian Strategy in Iraq: Politics and "Other Means,"* Occasional Paper Series, U.S. Military Academy, West Point, N.Y.: Combating Terrorism Center, October 13, 2008.

sector fundraising from only one of six sectors, and in that case, we have it for a time period just after the data from the Anbar provincial administration ended.

We analyze the associations between attacks and spending sector by sector. AQI in Anbar subdivided the province into six sectors, but the SIGACTS data on attacks combined two of them, ar-Ramadi and at-Ta'mim, so we combined spending in these sectors as well, with the result that our final data set subdivides Anbar into only five sectors.

We first show the relationship on a monthly basis and then move to a weekly analysis in which we estimate the relationship between attacks and spending in the weeks before attacks occur. The results show how expenditure changes are related to the change in the pace of attacks and provide an implicit estimate of how interdicting or disrupting insurgent financial networks would decrease the level of threat.

We find that a transfer of $2,700 from the Anbar administrative emir to a specific sector is related to an additional attack in that sector. We then put this figure in the context of more general findings about the costs of attacks and the costs of running a militant group. We conclude with caveats about the various potential sources of error in our analysis.

Monthly Patterns

Monthly attack levels in Anbar appear to be strongly related to monthly transfers to the sectors, with spending and attacks following the same general trend (Figure 5.1). In fact, the correlation between spending and attacks is 0.66. A simple regression of monthly attacks in Anbar on monthly spending in Anbar shows that this relationship is statistically significant at the 0.01 level or better, meaning that there is less than a one in 100 chance that the relationship is merely a random occurrence.

To test this relationship more formally and use the power of the sector-level data, we regressed the previously identified outcome attack variable from the SIGACTS data on monthly expenditures by sector. Equation 5.1 shows the estimating equation.

Figure 5.1
Monthly Spending and Attacks in Anbar, 2005–2006

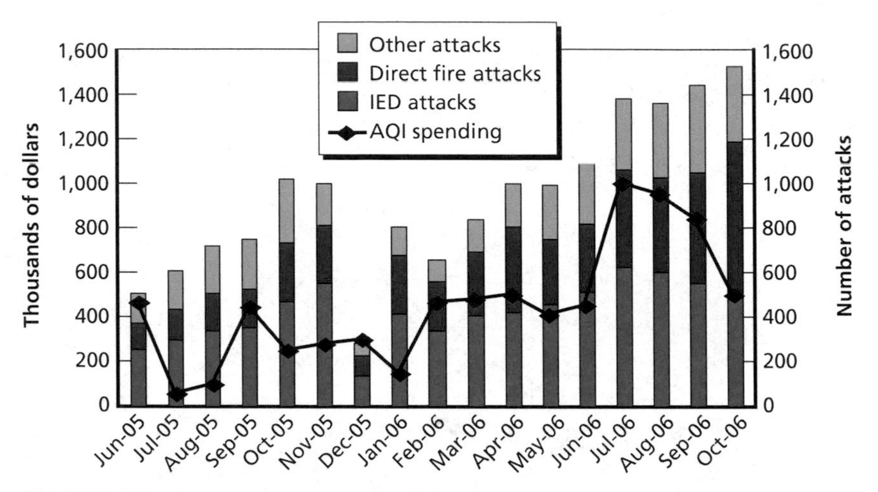

SOURCES: SIGACTS III and Harmony Batch ALA DAHAM HANUSH.

NOTES: AQI spending figures exclude the large transfers to the General Treasury in March, July, and August 2006, as we believe that these funds were spent outside Anbar province. The number of attacks was extraordinarily low in December 2005 because of the tribes' call for a hiatus in violence during national elections. In addition to total attacks, the SIGACTS data provide information on the type of attack that occurred: direct fire (DF) attacks, IED attacks, and other attacks. DF and IED attacks were the two most common types of attacks during the period. DF attacks include attacks by rocket-propelled grenade, snipers, drive-by shootings, and small arms. IED attacks include not just IEDs that exploded but also vehicle-borne IEDs (VBIEDs), suspected IEDs, and IEDs and VBIEDs that were cleared before they could be used.

RAND *MG1026-5.1*

$$A_{sm} = \alpha + \beta_1 E_{sm} + \beta_s I_s + \varepsilon_{sm} \tag{5.1}$$

In this equation, A_{sm} stands for attacks in each sector in each month of the data, E_{sm} stands for expenditures in each sector in each month, I_s is an indicator variable for each sector and is equal to 1 if the particular observation is attacks and expenditures in that sector, and 0 otherwise, and ε_{sm} is a random error term. Alpha is a constant, and each beta is a coefficient showing the relationship of each variable to attacks. The indicator variable accounts for the many unobserved variables that are constant within a sector during the time period and that may also be related to the number of attacks. In the analysis, spending is in thousands of dollars, and attacks are a simple count.

The results show that monthly spending in a sector is strongly associated with the number of attacks both substantively and statistically (Table 5.1). Every $1,000 spent is related to a bit more than 0.8 overall attacks, amounting to an average expenditure of $1,233 per attack. The relationship is also strongly statistically significant, indicating that this is not simply a random relationship.

Weekly Patterns

The monthly results indicate a strong relationship between spending and attacks, but the data allow us to be far more exact in understanding the timing between expenditures and attacks. Specifically, we now explore weekly expenditures and attacks, again at the sector level within AQI's Anbar organization. Using weeks as the unit of time increases the power of our analysis by giving four times the observations in the time dimension. A weekly level of analysis also shows the association of spending over several weeks to produce an attack.

Table 5.1
Relationship Between Monthly Attacks and Monthly Expenditures

Dependent Variable	All Attacks
Spending coeficient	0.811
Standard error	0.155
Overall R^2	0.3537
N	85

SOURCES: Authors' analysis of data from SIGACTS III and Harmony Batch ALA DAHAM HANUSH.

NOTE: The spending coefficient is significant at the 0.01 level or better and shows the relationship between a change in attacks and a change in spending when spending is changed by $1,000.

At a weekly level, spending by AQI's Anbar administration is correlated with attacks, although less so than at a monthly level (Table 5.2). In addition, this correlation extends through time across the previous month's spending. It is interesting to note that there is only a weak correlation from one week to the next, but spending in alternate weeks is somewhat stronger. We believe that this may indicate a tendency to make transfers semi-weekly.

Estimating by week allows us to include lagged spending to find the relationship, for example, between changes in spending three weeks ago and an attack this week. However, estimating the effect of spending on attacks without taking account of other events could lead to finding a spurious relationship. The range of factors, both observable and unobservable, that could influence the pace of attacks is potentially very large. We take account of these factors that are missing from our estimates by using indicator variables and time trend variables.

Table 5.2
Correlations Between Attacks and Spending

		Spending				
		Number of Weeks Before Attack				
	Attacks	0	1	2	3	4
Attack variables						
All attacks	1.00					
Spending variables						
Same week (0)	0.41	1.00				
One week before (1)	0.39	0.08	1.00			
Two weeks before (2)	0.41	0.27	0.08	1.00		
Three weeks before (3)	0.38	0.18	0.27	0.08	1.00	
Four weeks before (4)	0.30	0.07	0.12	0.22	0.08	1.00

SOURCES: Authors' analysis of data from SIGACTS III and Harmony Batch ALA DAHAM HANUSH.

NOTES: Attack and spending data are aggregated at the weekly level by sector. The data include 72 weeks and five sectors. There are only 340 total observations because of missing data.

Specifically, we include an indicator for each sector of Anbar province. This accounts for specific, though unchanging, characteristics of each sector that influenced the level of attacks. We include a variable for a time trend and a time trend squared and interact them with the sector indicators. These variables account for conditions that changed gradually, either in a straight-line fashion or by accelerating or decelerating, throughout the period in both Anbar province as a whole and in each sector. Finally, we include an indicator variable for each five-week period in the data. This variable takes account of specific events that happened approximately month by month that might have influenced the level of attacks. In unreported results, we also tried a specification in which we interacted the five-week indicators with the sector indicators. We discuss those results after reporting the main results.

With that background, our estimating equation is shown in Equation 5.2.

$$\begin{aligned} A_{sw} &= \alpha + \beta_1 E_{sw} + \beta_2 E_{sw-1} + \beta_3 E_{sw-2} + \beta_4 E_{sw-3} \\ &\quad + \beta_5 E_{sw-4} + \beta_{6s} I_s + \beta_7 T + \beta_8 T^2 + \beta_{9s} I_s * T \\ &\quad + \beta_{10s} I_s * T^2 + \beta_{11w} W5_{w5} + \varepsilon_{sw} \qquad (5.2) \end{aligned}$$

In this equation, A_{sw} is the number of attacks by sector by week, and E is expenditures by sector by week in the week of the attacks, the week before the attacks, two weeks before the attacks, three weeks before the attacks, and four weeks before the attacks. I is the indicator variable for each sector, T is a time trend, T^2 is the square of a time trend, $I*T$ is the interaction of the sector indicator and the time trend, $I*T^2$ is the interaction of the sector indicator and the square of the time trend, $W5$ is the indicator variable for every five-week period, and ε_{sw} is a random error term. As before, alpha is a constant and each beta is a coefficient showing the relationship of each variable to attacks.

We estimate the equation in two ways. First, we use just the expenditure variables and the sector indictor. This is shown as Column A in Table 5.3. Second, we include all variables in Equation 5.2. This is our preferred estimate, since it accounts for more

Table 5.3
Relationship Between Weekly Expenditures and Weekly Attacks by AQI in Anbar, June 2005 Through October 2006

	Attacks	
Specification	**A**	**B**
Spending same week		
Coefficient	0.218**	0.100**
Standard error	0.044	0.033
Spending one week before		
Coefficient	0.218**	0.083*
Standard error	0.044	0.034
Spending two weeks before		
Coefficient	0.230**	0.084*
Standard error	0.045	0.034
Spending three weeks before		
Coefficient	0.182**	0.037
Standard error	0.044	0.035
Spending four weeks before		
Coefficient	0.137	0.062*
Standard error	0.037	0.030
Overall R^2	0.4552	0.805
N	340	340

SOURCES: Authors' analysis of data from SIGACTS III and Harmony Batch ALA DAHAM HANUSH.

NOTES: The Column A estimation includes an indicator variable for each sector. The Column B estimation includes the sector indicator, a time trend and time trend squared, both interacted with the sector indicator, and an indicator for every five-week period.

* Indicates statistical significance at the 0.05 level, meaning that there is at least a 95 percent chance that the relationship as estimated is different from zero; put another way, this means there is at least a 95 percent chance that a relationship between spending and attacks exists.

** Indicates statistical significance at the 0.01 level, meaning that there is at least a 99 percent chance that the relationship as estimated is different from zero; put another way, this means there is at least a 99 percent chance that a relationship between spending and attacks exists.

observable and unobservable variables that could influence attacks, and is shown as Column B in Table 5.3.

Sector-specific spending by the AQI administrative emir in Anbar shows a statistically significant association with the number of attacks in that sector (with a p-value of less than 0.01, or $p < 0.01$) across many weeks of spending before the attack. In the specification of Column A, spending has a statistically significant association with the number of attacks even three weeks into the future. The coefficient for spending has comparable magnitude over time, indicating that sector spending is paying for attacks throughout the coming month.

However, as noted above, the specification in Column A could omit other variables related to the number of attacks. The results shown in Column B indicate that this is indeed a problem. But even in the Column B specification, our preferred specification, total attacks have a statistically significant association with spending in the same week ($p < 0.01$), one week prior ($p < 0.05$), two weeks prior ($p < 0.05$), and four weeks prior ($p < 0.05$). These associations show a much smaller effect of AQI in Anbar's spending than the specification in Column A showed; the trend variables and other indicators tend to be positively correlated with both spending and attacks. This would be the case if the interacted trend variables for the sectors were capturing spending of the individual sectors, and if specific events during each five-week period caused sector-level spending to rise or fall. Such a relationship is almost certain, given that the pace of AQI attacks responded to the pace of counterinsurgency and counterterrorism operations by CF.

The fact that spending in one period is associated with future attacks may indicate that AQI conducts extensive planning and funding throughout the lead-up to an attack. Although our model does not test for this kind of difference, costs at different stages before the attack may correspond to different types of costs (such as procurement versus labor). This result further strengthens the argument that AQI administrators, who collect, oversee, store, and distribute funds, are of great importance to the attack cycle, and that one-time payments are not sufficient to conduct either simple or complex attacks.

Changes in Spending and Changes in Attacks

Using specification B in Table 5.3, we find that for every $2,732 transmitted by the Anbar administrative emir to a particular sector, an additional attack occurred in that sector (Table 5.4). This estimate is greater than that based on specification A of Table 5.3 ($1,015) or the estimate based on the monthly unit of analysis shown in Table 5.1 ($1,233), but the $2,732 figure is by far our preferred estimate. The additional controls for unobserved variables make this estimate the most accurate. These coefficients for expenditures as a group are statistically significant ($p < 0.01$).

We compute the $2,732 figure as follows. Each spending coefficient shows how much a $1,000 change in spending in a given week would change the number of attacks. Using the coefficients in Table 5.3, increasing expenditures by $1,000 in a week would increase the number of attacks by 0.1 in the same week, by 0.08 one week later, by 0.08 two weeks later, by 0.04 three weeks later, and by 0.06 four weeks later, for a total of almost 0.4 additional attacks. To convert this to the additional expenditure needed for one complete attack, we divide $1,000 by 0.366 (the exact fractional increase in attacks) to get $2,732.

The higher cost found by the addition of many control variables indicates that more variables than just spending influence the level of attacks. However, even with these control variables included, our estimates may still be inaccurate for a number of reasons. We detail the caveats to the analysis at the end of the chapter. In the next section, we put our findings in broader context.

Table 5.4
Estimates of Changes in Spending and Attacks

	Weekly		Monthly
Specification	A	B	
Dollars per attack	1,015	2,732	1,233

SOURCES: Authors' analysis of data from SIGACTS III and Harmony Batch ALA DAHAM HANUSH.

NOTES: Weekly specifications appear in Table 5.3. The monthly specification appears in Table 5.1.

The Cost of Militant Activity

Our findings shed light on the debate over the cost of running a militant, terrorist, or insurgent organization. It is recognized that the cost of the actual equipment used in an attack can be quite low. For example, the ingredients used to build each bomb intended to blow up airliners bound for the United States from the United Kingdom in 2006 are estimated to have cost only $15.[2] The cost of an IED has been estimated to be $25 to $30.[3] Similarly, the material cost for conducting a suicide bomb has been estimated at only $150.[4]

How then can it be that a transfer of more than $2,700 is associated with only one additional attack? The answer is that it takes more than just equipment for a militant group to conduct an attack. Referring to terrorist groups, the multilateral Financial Action Task Force (FATF) notes that

> The costs associated not only with conducting terrorist attacks but also with developing and maintaining a terrorist organization and its ideology are significant. Funds are required to promote a militant ideology, pay operatives and their families, arrange for travel, train new members, forge documents, pay bribes, acquire weapons, and stage attacks. Often, a variety of higher-cost services, including propaganda and ostensibly legitimate social or charitable activities are needed to provide a veil of legitimacy for organizations that promote their objectives through terrorism.[5]

The same holds for militant groups more generally.

For example, the FATF estimated that the bombings of two U.S. embassies in East Africa in 1998 had direct costs of $50,000. How-

---

[2] Craig Whitlock, "Al-Qaeda Masters Terrorism on the Cheap," *The Washington Post,* August 24, 2008.

[3] David Axe, "Soldiers, Marines Team Up in 'Trailblazer' Patrols," *National Defense: NDIA's Business and Technology Magazine,* April 2006.

[4] Bjorn Lomborg, "Is Counterterrorism Good Value for Money?" *The Mechanics of Terrorism, NATO Review,* April 2008.

[5] Financial Action Task Force—Groupe d'Action Financière, *Terrorist Financing,* Paris, France: FATF Secretariat, 2008, p. 7.

ever, it differentiated between direct costs and broader organizational costs. Prober (2005) estimates that the broader organizational costs were much higher and likely included

- setting up and running al-Qa'ida–affiliated businesses
- travel for senior al-Qa'ida members to Nairobi
- training
- the rental of an estate and its use as a bomb factory
- maintaining communications, including the use of expensive satellite telephones
- bribes.[6]

Despite the recognition of organizational versus direct costs, there is still disagreement about whether militant group actions are inexpensive or expensive.[7] However, the definitions of expensive and inexpensive may be in the eye of the beholder. Our analysis provides a more concrete answer. The amount $2,700 is equivalent to almost three times Anbari per capita 2007 household income (in 2006 dollars) and 40 percent of total average household income, a relatively large sum.

What we do know is that a transfer of $2,700 did not lead to 18 more suicide bombs, which would have been the case if a suicide bomb cost only $150. Nor did it lead to 90 more IEDs, which would have been the case if IEDs cost only $30 per attack. Rather, AQI not only had to buy equipment for attacks, but it had to prepare the operation and absorb the broader organizational costs of running an insurgency. Preparing the operation could include engaging in reconnaissance, identifying points that would have the greatest payoff in terms of damage to the target, and balancing that against points that would give AQI mili-

---

[6] Joshua Prober, "Accounting for Terror: Debunking the Paradigm of Inexpensive Terrorism," Policy Watch #1041, Washington, D.C.: The Washington Institute for Near East Policy, November 1, 2005.

[7] On the Iraq insurgency, see Steven Metz and Raymond Millen, "Insurgency in Iraq and Afghanistan: Change and Continuity," prepared for the National Intelligence Council, Carlisle, Pa.: Strategic Studies Institute, U.S. Army War College, 2004; on terrorism more generally, see Greg Bruno, "Al-Qaeda's Financial Pressures," *Backgrounder*, Council on Foreign Relations, February 1, 2010.

tants the best change of escaping capture. The broader organizational costs for AQI entailed not just compensation, rents, facilities, vehicles, and medical expenses, but, also in the case of expanding the pace of attacks, adding members to units or adding entirely new units. This provides added evidence that, to the extent that the broader costs are needed to sustain militant group activity, interdicting finances could degrade group activities and lower both the quantity and effectiveness of attacks.

Caveats to the Analysis

Testing statistical associations between the level of attacks in a sector and the AQI Anbar administrative emir's transfers of money to the sector rests on a number of assumptions that may introduce biases into the analysis. Some of these assumptions are quite strong. We have tried to mitigate some of these potential biases by adding indicator and trend variables. Most, but not all, of these problems bias our cost estimate downward, and so we believe our estimate should be treated as a lower bound of AQI's cost per attack in Anbar in 2005 through 2006.

The analysis assumes that most, if not all, of the attacks in the SIGACTS database in Anbar are perpetrated by AQI or that attacks in the database perpetrated by AQI are uncorrelated with attacks in the database perpetrated by other groups. In fact, unpublished RAND analysis linking SIGACTS data to Internet attack claims showed that more than 50 percent of attacks in Anbar during this time period were claimed by AQI. By attributing all SIGACTS attacks to AQI, we underestimate the cost per attack by the unknown proportion of attacks not conducted by AQI.

The analysis makes a number of implicit assumptions about financing behavior by AQI. By focusing only on transfers from the AQI Anbar administrator to the sectors, it omits additional spending by the sector-level components of AQI or other AQI spending in support of Anbar operations. We know from the Tuzliyah documents that the Anbar administrative emir's transfers form a large part of sector funding. The western sector funded one-third of its spending with transfers from the Anbar provincial administrative emir in late 2006.[8] Assuming that the western sector is representative of the six sectors, the transfers of the Anbar emir to each sector should account for roughly half as much money as the spending that stemmed from revenues originating in each sector. The possibility that sectors self-fund attacks in addition to using transfers from the provincial administrative emir for attacks biases our estimate of the cost per attack downward. Alternatively, if sector spending is inverse to transfers (the sectors self-fund more when transfers are low, or self-fund less when transfers are high), then our estimates may be about right or even high.

The analysis also assumes that funding not directed to a specific sector does not correspond with an increased level of attacks. We do not believe this is a problem, as all evidence in the documents points toward sectors as being the primary bureaucratic level in charge of the units that conduct attacks.[9] However, spending by the Anbar provincial administration that was not directed to specific sectors may in fact have provided them services that helped them increase their level of attacks. Nonsector-specific spending accounts for 41.5 percent of total AQI Anbar expenditures. If this spending provides support for attacks, then our cost estimates are once again too low.

In at least one case, our estimates might be biased upward. Although SIGACTS is reported to contain attacks not just on CF, but also on ISF, civilians, and Iraqi infrastructure, the latter three attacks appear to be underreported. If this is the case, then our cost figures are an overestimate.

Our analysis also assumes that attacks are determined by spending alone and that spending is not related to the level of previous attacks. We have responded to the problem of omitted variables by including indicator and time trend variables to account for events and trends that are not in our data set. This also helps account for previ-

---

8 Harmony Batch MA7029-5.

9 In a regression replicating Table 5.3, column B, for all attacks and including all nonsector spending, the coefficient for nonsector spending was not statistically significant, and the other coefficients were almost exactly the same in both magnitude and statistical significance as in the table.

ous attacks.[10] We performed an even stronger test, not reported, in which we ran both specifications A and B of Table 5.3 and included sector-specific, five-week indicator variables. These would account for any event—including spending—that occurred in a sector in a specific five-week period. In both cases, none of the spending variables were statistically significant. The point estimates for specification A result in a cost per attack of $6,329, and the point estimates of specification B result in a cost per attack of $15,151. Because the sector-specific five-week indicators could proxy for spending, we do not consider that these results invalidate our previous findings. However, they do suggest that events other than spending also influenced the number of attacks.

Given these caveats and the ways we have tried to counter any apparent problems with the data or the estimation techniques, we believe that the estimate of $2,732 roughly corresponds to the true relationship between transfers from the AQI Anbar administrator to a sector and an additional attack in that sector. In addition to buying the materiel for attacks, AQI has extensive recurring expenses for administration and payroll. Attacks are expensive, and maintaining the level of violence practiced by AQI in Anbar during 2005 and 2006 required substantial financial resources.

---

[10] We chose not to enter lagged attacks directly in the estimating equations because using lagged dependent variables in a panel structure presents a number of econometric problems that make the estimates not wholly reliable. We plan for future research to include these more complex specifications.

CHAPTER SIX

Implications

The data in these sets of documents reveal the practical, day-to-day financial activity of an al-Qa'ida–affiliated group at the height of its power in a specific place—Anbar province, Iraq—and a specific time—2005 and 2006. Our findings expose patterns of behavior that Iraqi and U.S. counterterrorism authorities may be able to exploit against AQI and other insurgent groups. Although our findings do not necessarily apply to the broader range of salafi jihadist militant groups or even to al-Qa'ida as a whole, the data provide a unique look inside a radical militant organization that is, as of the end of 2010, still attempting to subvert and destroy the existing Iraqi government to establish an Islamic state.

One important implication for all current and future counterterrorism and counterinsurgency operations is that a militant group's financial records may hold valuable information about the group's operations and motivations. At least in the period covered by this set of documents, senior financial administrators kept detailed records, including the names—albeit in alias—of operatives and deceased members, along with all financial flows, equipment purchases, and other types of information that can inform U.S. or partner nation operations. In general, these records often survive a raid, and, if properly preserved, documented, and exploited in a timely manner, provide valuable intelligence on the groups' command and control, funding, and decisionmaking.

Implications About Organization and Financing Methods

A number of implications can be drawn from the analysis in the preceding chapters. First, the documents revealed that AQI's hierarchical, bureaucratic structure in Anbar provides a set of rules that helps to ensure that resources are shared across geographic and functional areas. Funding flows from the group's central administration to strategically important operational units, and the central provincial administration is able to collect revenues from lucrative areas operated by subsidiary units. Collecting funds, then transmitting them to and from different administrative units and eventually back to operational cells, presents a potential vulnerability. The constant movement of cash, presumably by courier, gives security forces more opportunities to penetrate AQI's financial system. However, bulk cash is easy to conceal, and creating chains of couriers can protect the administrative emirs from being tracked down easily.

Large amounts of cash passed through the administrative emirs' hands, but it moved quickly, offering only a short window for CF efforts to disrupt the flow to attackers in the field. However, the fact that the group appears to transfer funds solely through cash couriers may indicate the success of U.S. and Iraqi authorities in attempting to deter militants from using formal and informal financial institutions because of a perceived risk of interdiction.

Delegating the decisionmaking about expenditures to the local level provided flexibility and adaptability for local group leaders, allowing them to maintain operational tempo despite short-term disruptions in the command chain (for example, the death of Zarqawi in 2006). However, decentralized decisionmaking can also be a vulnerability. It increases the ability of local leaders to avoid monitoring by higher authorities, and this can lead to activities that run counter to the interest of the organization as a whole if the incentives of the different organizational units are not aligned. For example, a sector that is particularly aggressive about kidnappings and extortion could alienate the local population when it is in the interests of the group to avoid doing so. Decentralization also suggests that central government or other forces fighting the insurgent group have a greater opportunity to make side deals to gain informants or to negotiate with fringe elements of the group.

Where efforts to counter other militant groups' funding often have focused on major donors, the revenue streams of AQI recorded in these documents consisted mostly of stolen money, equipment, and retail goods, as well as the sale of cars and spoils—a more diversified and thus more difficult revenue stream to disrupt. However, these revenues open the group to the criticism that they are criminally minded rather than ideological. In this case, the political economy of the group's revenue—in terms of the various revenue sources and AQI's impingement on tribal profits—may have been a key strategic blunder that galvanized a number of tribal sheiks to form the Awakening movement, aside from AQI's murder and intimidation campaign.

This key lesson from AQI's history may provide a strategic approach for security forces in Afghanistan and other theaters of protracted low-intensity warfare. In funding themselves, militant groups create losers from other players in the economy. In the case of Anbar province, AQI activities cut into the economic activity of tribal networks, providing a spur to the tribes to organize opposition. In a case such as Afghanistan, security forces and policymakers may have to be more creative in organizing the opposition of the economically displaced, but there are likely to be such groups that could be incentivized to move against the Taliban or al-Qa'ida. However, it must be acknowledged that difficult trade-offs are inherent to arming and supporting such groups to fight militants, particularly in the face of an ineffectual and weak central government. This was hotly debated surrounding CF support of the Awakening movement, and a similar debate is ongoing in regard to supporting tribal militias in Afghanistan.

Implications About Spending and Attacks

Increased spending from the AQI Anbar administration to its sectors increases the number of attacks in those sectors, with one additional attack occurring for every additional $2,700 transferred. Increases in spending appear to be associated with an increase in attacks in the same week of spending as well as in future weeks. This suggests that financial interdiction, although difficult, can slow the pace of insurgent activity.

Putting together an IED or buying a mortar for an attack is cheap. However, our findings add to the mounting evidence that militant group operations involve far more than just one-time costs. Maintaining a militant organization can be quite expensive. For AQI, personnel costs for members constituted the bulk of these expenses. Without such recurring payments, it is unlikely that AQI could maintain its effectiveness in committing violence. The group incurred large costs keeping imprisoned members on the payroll as an obligation to their families and paying the families of dead members. Although such payments likely increased the loyalty of members, they also diverted large amounts of money that could have otherwise been used to attack Coalition and Iraqi forces.

Compensation as a Guide to Motivations

Personnel and pay records are crucially important in estimating the financial benefits of group membership, as well as in understanding the amount of physical risk being imposed on the membership by security force activities. Such records can also shed light on motivations beyond the simple desire for compensation.

AQI members in Anbar province received compensation below that received by the general population, and they also faced a much higher risk of death. This result shows that if AQI members were economically rational, financial rewards were not among their primary motivations for joining the group. Furthermore, it is highly likely tht AQI members knew that they were giving up the equivalent of many years of income by joining the group. Combined with our findings about spending and attacks, this indicates that although disrupting the group's finances could have a favorable affect on degrading its ability to sustain high levels of attacks, it may not have a strong effect on motivations for joining or remaining a member. Consequently, this study validates the idea that financial interdiction can serve as one tool in counterinsurgency but is unlikely on its own to bring about the demise of a militant group.

Threat Finance or Threat Economics?

The analysis provided in this monograph supports and potentially strengthens the notion that militant finances should be tracked by U.S. and partner nation intelligence, but our findings also suggest that the notion of threat "finance" may be too narrow. If commanders and policymakers are informed only of the revenue side, half of the useful data on militant back office operations may be ignored. As we have shown, understanding the spending patterns of militant groups may be just as useful as understanding their revenue streams. Although the U.S. Department of Defense and the U.S. intelligence community have wisely expanded their resources on the topic of militant finances, these findings suggest that a similar expansion into following militant expenditures is warranted. We therefore recommend that the concept of "threat finance" be broadened to "threat economics" to better frame the attention of analysts, operators, commanders, and policymakers.

APPENDIX A

Anbar Province

Figure A.1
Map of Districts and Selected Cities of Anbar Province

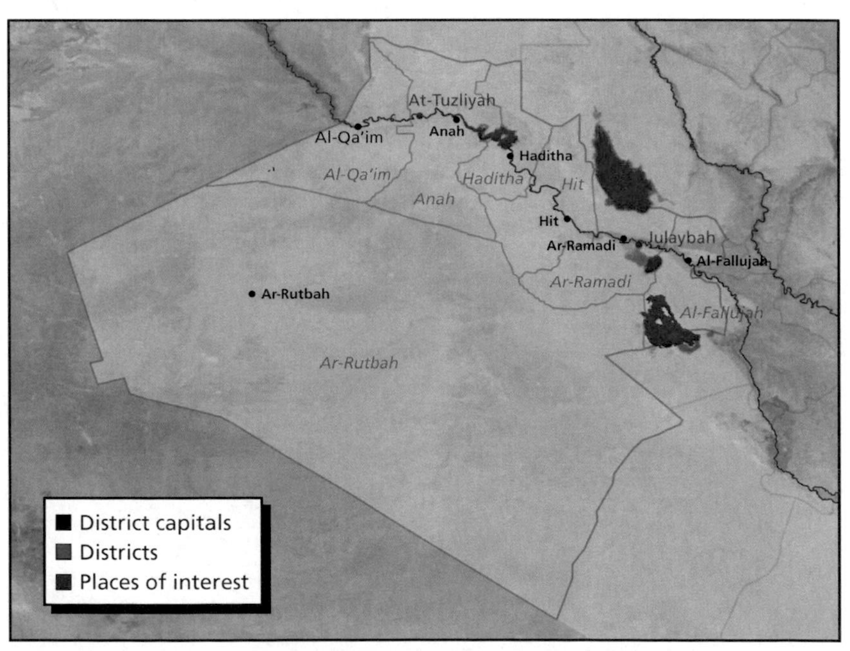

SOURCES: Harmony Batches MA7029-5 and ALA DAHAM HANUSH.
NOTE: The image shows districts, cities, towns, and other geographic landmarks in Anbar province.
RAND *MG1026-A.1*

Figure A.2
Map of AQI Sectors in Anbar Province

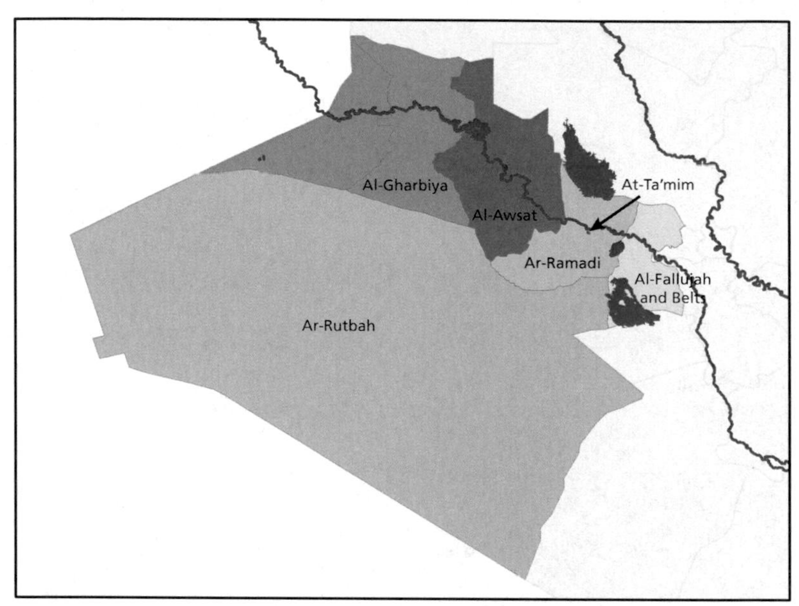

SOURCES: Harmony Batches MA7029-5 and ALA DAHAM HANUSH.
NOTES: The image shows the authors' assessments of AQI's six sector divisions of Anbar province. These six include ar-Rutbah, al-Gharbiyah, al-Awsat, at-Ta'mim, ar-Ramadi, and Fallujah and belts. These divisions are similar to the official Iraqi administrative divisions of Anbar province.

APPENDIX B

Time Line of Events in Anbar Province

Table B.1
Time Line

Date	Event	Comment
Circa 1999	Jund al-Sham (Army of Greater Syria) founded by Abu Mus'ab al-Zarqawi in Afghanistan	Specific details unconfirmed; focused on anti-Western, anti-Jordanian, and anti-Israeli operations in the Levant
March 2003	Coalition Forces invade Iraq	Much of Saddam's armed forces do not stand and fight
Early 2004	Operation Vigilant Resolve in Fallujah	Largely considered a failure as it provides insurgents a publicity win and does not clear Fallujah of extremists
Late 2003 or April 2004	Zarqawi establishes Jama'at al-Tawhid wa al-Jihad (Monotheism and Jihad Group); his forces build strength in Anbar	Goal to force Coalition Forces, foreign companies out of Iraq; push Iraqis to stop supporting U.S. and Iraqi governments
October 17, 2004	Zarqawi swears allegiance to Usama Bin Ladin and merges JTJ with al-Qa'ida "central" to form al-Qa'ida in Iraq	Similar to JTJ, but tries to make Iraq the center of a greater Islamic state by also spreading the jihad into "Greater Syria"
November 2004	Coalition Forces clear Fallujah of AQI in Operation al-Fajr	Widely considered a success due to tribal buy-in, establishment of a police force, and CF civil affairs work
January 2005	Tribes boycott provincial elections in Anbar	Only 3,700 votes tallied across the province

Table B.1—Continued

Date	Event	Comment
May 2005	CF begin extensive operations along Syrian border with Anbar[a]	
June 2005	First date of available financial records by AQI Anbar province administrator "Firas"	
July 9, 2005	Bin Ladin deputy Ayman al-Zawahiri pens letter to Zarqawi chastising him for his tactics in publicizing brutal attacks on Iraqi civilians	Letter was not publicly available until October 2005
August 2005	Attempted AQI rocket attack on U.S. Navy ship in Aqaba, Jordan	
September 2005	Abu Azzam killed—"main gatekeeper" for AQI, with operational control over flow of money from Baghdad to rest of Iraq	
Fall of 2005	Heavy fighting between Albu Mahal and Dulaimi tribes with AQI in al-Qa'im and Ramadi	
November 2005	Operation Steel Curtain clears al-Qa'im	Joint operation with Albu Mahal tribe
November 9, 2005	AQI attacks three hotels in Amman, Jordan, killing more than 50	
Mid-December 2005	Successful national elections; establishment of Anbar People's Council; temporary pause in violence	
January 2006	Increase in ISF recruitment in Anbar; AQI creates the Mujahidin Shura Council as an umbrella organization to encompass AQI and five other groups	Zarqawi raises priority on targeting Shi'a civilians
End of January 2006	Dramatic increase in violence; Anbar People's Council leaders largely eliminated by AQI	

Table B.1—Continued

Date	Event	Comment
March and April 2006	Plans drawn up to clear Ramadi without tribal support	MNF-W increases force levels in Ramadi
May 2006	AQI Anbar province financial administrator "Firas" replaced by "'Imad"	U.S. military touted successes in Iraq in May 4, 2006, briefing and released captured photos of Zarqawi
June 7, 2006	Zarqawi killed in U.S. airstrike	Abu Ayyub al-Masri (a.k.a. Abu Hamza al-Muhajir) takes over AQI
August–October 2006	Sheik Sattar Abu Risha and other activist tribal leaders around Ramadi organize the Awakening movement	Goal to fight AQI and like-minded extremists
October 15, 2006	AQI and MSC, with other groups, declare formation of Islamic State of Iraq under Abu 'Umar al-Baghdadi; showcase size of territorial control[b]	In addition to above AQI and MSC goals, aims to put an Iraqi face on the movement and lay groundwork for the caliphate by sending out emirs to enforce its writ in various parts of Iraq
October 16, 2006	'Imad eliminates three sectors of Anbar province in records and reorganizes Ramadi, Fallujah, and "Central" sectors	
November 2006	Last period of available Anbar provincial administration financial records	
January 2007	President George W. Bush announces the Baghdad security plan, otherwise known as "the Surge"	Plan extends the tours of a few units in Ramadi, increasing CF force levels
January–April 2007	Anbar capital Ramadi gradually cleared of ISI insurgents	Police recruitment increases, AQI conducts counterattacks
March 15, 2007	ISI Anbar financial documents captured in Julaybah raid	
Mid-2007	Zobai tribe (part of Shammari confederation) turns against AQI in most of its tribal area in Anbar[c]	According to Kilcullen, contradicted AQI claim that it was only thing standing between Sunnis and Shi'a-led genocide[d]
Early 2008	Sheik Sattar Abu Risha assassinated by AQI	

Table B.1—Continued

Date	Event	Comment
September 19, 2008	Abu Hamza al-Muhajir lecture "The State of the Prophet" attempts to justify ISI, yet says that territorial control is not the correct yardstick to measure legitimacy or strength[e]	

[a] Mahan Abedin, "Anbar Province and Emerging Trends in the Iraqi Insurgency," *Jamestown Foundation Terrorism Monitor,* Vol. III, Issue 14, July 15, 2005.

[b] Pascale Combelles Siegel, "Islamic State of Iraq Commemorates Its Two-Year Aniversary," *CTC Sentinel,* Vol. 1, Issue 11, October 2008, p. 6.

[c] Kilcullen, 2008, pp. 1–5.

[d] Kilcullen, 2008, p. 4.

[e] Combelles Siegel, 2008, p. 6.

Bibliography

Abedin, Mahan, "Anbar Province and Emerging Trends in the Iraqi Insurgency," *Jamestown Foundation Terrorism Monitor,* Vol. III, Issue 14, July 15, 2005.

Adams, James, *The Financing of Terror,* New York: Simon & Schuster Press, 1986.

"Al-Furqan Media Wing Declares the Members of the Cabinet of the Islamic State of Iraq," April 19, 2007, as cited in Evan F. Kohlmann, "State of the Sunni Insurgency in Iraq: August 2007," New York: The NEFA Foundation, 2007. As of July 25, 2010:
http://www.nefafoundation.org

Axe, David, "Soldiers, Marines Team Up in 'Trailblazer' Patrols," *National Defense: NDIA's Business and Technology Magazine,* April 2006.

Benmelech, Efraim, and Claude Berrebi, "Attack Assignments in Terror Organizations and the Productivity of Suicide Bombers," National Bureau of Economic Research Working Paper W12910, Cambridge, Mass., 2007a.

———, "Human Capital and the Productivity of Suicide Bombers," *The Journal of Economic Perspectives,* Vol. 21, No. 3, Summer 2007b, pp. 223–238.

Berman, Eli, "Hamas, Taliban and the Jewish Underground: An Economist's View of the Radical Religious Militias," National Bureau of Economic Research Working Paper W10004, Cambridge, Mass., 2003.

Berman, Eli, and David D. Laitin, "Religion, Terrorism and Public Goods: Testing the Club Model," *Journal of Public Economics,* Vol. 92, Nos. 10–11, 2008, pp. 1942–1967.

Berman, Eli, Jacob N. Shapiro, and Joseph Felter, "Can Hearts and Minds Be Bought? The Economics of Counterinsurgency in Iraq," National Bureau of Economic Research Working Paper W14606, Cambridge, Mass., 2008.

Berrebi, Claude, "Evidence About the Link Between Education, Poverty, and Terrorism Among Palestinians," *Peace Economics, Peace Science, and Public Policy,* Vol. 13, No. 1, 2007, pp. 1–36.

———, "The Economics of Terrorism and Counterterrorism: What Matters and Is Rational-Choice Theory Helpful?" in Paul K. Davis and Kim Cragin, eds., *Social Science for Counterterrorism: Putting the Pieces Together,* Santa Monica, Calif.: RAND Corporation, MG-849-OSD, 2009. As of June 29, 2010: http://www.rand.org/pubs/monographs/MG849/

Bruno, Greg, "Al-Qaeda's Financial Pressures," *Backgrounder,* Council on Foreign Relations, February 1, 2010.

Burns, John F., and Kirk Semple, "U.S. Finds Insurgency Has Funds to Sustain Itself," *New York Times,* November 26, 2006.

Caplin, Andrew, and John Leahy, "The Social Discount Rate," *The Journal of Political Economy,* Vol. 112, No. 6, December 2004, pp. 1257–1268.

Central Bank of Iraq, "Key Financial Indicators for February 17, 2010," Excel spreadsheet, Baghdad, Iraq, 2010.

Central Organization for Statistics and Information Technology, Iraq, Kurdistan Region Statistics Office, and International Bank for Reconstruction and Development/The World Bank, *Iraq Household Socio-Economic Survey IHSES-2007,* Baghdad, Iraq, 2008a.

Central Organization for Statistics and Information Technology, Iraq; Ministry of Planning and Development Cooperation, Iraq; Kurdistan Region Statistics Office, Iraq; Nutrition Research Institute, Ministry of Health, Iraq; and United Nations World Food Programme, Iraq Country Office, *Comprehensive Food Security & Vulnerability Analysis in Iraq 2007,* 2008b.

Combating Terrorism Center, Harmony Project Database, West Point, N.Y., undated. As of January 19, 2010: http://www.ctc.usma.edu/harmony/harmony_docs.asp

Combelles Siegel, Pascale, "Islamic State of Iraq Commemorates Its Two-Year Anniversary," *CTC Sentinel,* Vol. 1, Issue 11, October 2008.

Cooley, Alexander, *Logics of Hierarchy: The Organization of Empires, States and Military Occupations*, Ithaca, N.Y.: Cornell University Press, 2005.

Crane, Keith, Martin C. Libicki, Audra K. Grant, James B. Bruce, Omar Al-Shahery, Alireza Nader, and Suzanne Perry, *Living Conditions in Anbar Province in June 2008,* Santa Monica, Calif.: RAND Corporation, TR-715-MCIA, 2009. As of June 29, 2010: http://www.rand.org/pubs/technical_reports/TR715/

Cullison, Alan, "Inside Al-Qaeda's Hard Drive," *The Atlantic Monthly,* September 2004. As of January 3, 2009: http://www.theatlantic.com/doc/200409/cullison

Darwish, Adel, "Abu Musab al-Zarqawi," *The Independent,* June 9, 2006. As of January 19, 2010: http://www.independent.co.uk/news/obituaries/abu-musab-alzarqawi-481622.html

Ehrenfeld, Rachel, *Where Does the Money Go? A Study of the Palestinian Authority,* New York: American Center for Democracy, 2002. As of January 2, 2010: http://mefacts.org/cache/pdf/palestinians/10321.pdf

Eisenstadt, Michael, and Jeffrey White, "Assessing Iraq's Sunni Arab Insurgency," Washington Institute for Near East Policy, Policy Focus #50, December 2005. As of February 26, 2010:
http://www.washingtoninstitute.org/pubPDFs/PolicyFocus50.pdf

Fair, Christine C., and Bryan Shepherd, "Who Supports Terrorism? Evidence from Fourteen Muslim Countries," *Studies in Conflict and Terrorism,* Vol. 29, No. 1, 2006, pp. 51–74.

Farley, Jonathan D., "Breaking Al Qaeda Cells: A Mathematical Analysis of Counterterrorism Operations (A Guide for Risk Assessment and Decision Making)," *Studies in Conflict and Terrorism,* Vol. 26, 2003, pp. 399–411.

Felter, Joseph, and Brian Fishman, *Iranian Strategy in Iraq: Politics and "Other Means,"* Occasional Paper Series, U.S. Military Academy, West Point, N.Y.: Combating Terrorism Center, October 13, 2008.

Financial Action Task Force—Groupe d'Action Financière, *Terrorist Financing,* Paris, France: FATF Secretariat, 2008.

Fischer, Hannah, *Iraqi Civilian Deaths Estimates,* CRS Report for Congress RS22537, Washington, D.C.: The Library of Congress, August 27, 2008.

Fishman, Brian, "Dysfunction and Decline: Lessons Learned from Inside al-Qa'ida in Iraq," Harmony Project, U.S. Military Academy, West Point, N.Y.: Combating Terrorism Center, March 16, 2009.

———, ed., *Bombers, Bank Accounts, & Bleedout: Al-Qa'ida's Road In and Out of Iraq,* U.S. Military Academy, West Point, N.Y.: Combating Terrorism Center, 2008. As of January 2, 2009:
http://www.ctc.usma.edu/harmony/pdf/Sinjar_2_July_23.pdf

Freedman, Lawrence Zelic, and Yonah Alexander, eds., *Perspectives on Terrorism,* Wilmington, Del.: Scholarly Resources Inc., 1983.

Giraldo, Jeanne K., and Harold A. Trinkunas, eds., *Terrorism Financing and State Responses: A Comparative Perspective*, Stanford, Calif.: Stanford University Press, 2007a.

———, "The Political Economy of Terrorism Financing," in Jeanne K. Giraldo and Harold A. Trinkunas, eds., *Terrorism Financing and State Responses: A Comparative Perspective,* Stanford, Calif.: Stanford University Press, 2007b.

Hammes, Thomas X., "Countering Evolved Insurgent Networks," *Military Review,* July–August 2006, pp. 18–26.

Harmony Batch MA7029-5, documents MNFA-2007-000560, MNFA-2007-000562, MNFA-2007-000564, MNFA-2007-000566, MNFA-2007-000570, MNFA-2007-000572, MNFA-2007-000573, and MNFA-2007-000574.

Harmony Batch ALA DAHAM HANUSH, documents NMEC-2007-633541, NMEC-2007-633700, NMEC-2007-633893, and NMEC-2007-633919.

Harrison, Mark, "An Economist Looks at Suicide Terrorism," *World Economics,* Vol. 7, No. 3, 2006, pp. 1–15.

Hilsenrath, Peter E., and Karan P. Singh, "Palestinian Health Institutions: Finding a Way Forward After the Second Intifada," *Peace Economics, Peace Science and Public Policy,* Vol. 13, No. 1, Article 4, 2007.

Hudson, Rex A., *The Sociology and Psychology of Terrorism: Who Becomes a Terrorist and Why?* A Report Prepared Under an Interagency Agreement by the Federal Research Division, Library of Congress, Washington, D.C., September 1999. As of October 1, 2009:
http://purl.access.gpo.gov/GPO/LPS17114

Iannaccone, Laurence R., "Sacrifice and Stigma: Reducing Free-Riding in Cults, Communes, and Other Collectives," *The Journal of Political Economy,* Vol. 100, No. 2, 1992, pp. 271–291.

"Insurgency Is Raising Millions, Report Finds," *New York Times,* November 26, 2006.

Inter-Agency Information and Analysis Unit, "Anbar Governorate Profile," United Nations Office for the Coordination of Humanitarian Affairs, Amman and Baghdad, March 2009. As of February 26, 2010:
http://www.iauiraq.org/reports/GP-Anbar.pdf

International Crisis Group, "In their Own Words: Reading the Iraqi Insurgency," Amman and Brussels, 2006. As of January 3, 2010:
http://www.c4ads.org/files/ICG_report_021506_iraqi_insurgency.pdf

International Monetary Fund, "Iraq: Third and Fourth Reviews Under the Stand-By Arrangement, Financing Assurances Review, and Requests for Extension of the Arrangement and for Waiver of Nonobservance of a Performance Criterion," Washington, D.C., February 23, 2007.

Irabarren, Florencio, "ETA: Estrategia Organizative y Actuaciones, 1978–1992," Bilbao, Spain: Universidad del Pais Vasco, 1998, pp. 136–152.

Iraq Body Count Project, "IBC2006-ANBAR-DEMOG-4RAND.xls," Excel data file of noncombatant violent deaths in Anbar province, Iraq, prepared for the RAND Corporation, March 17, 2010.

Israeli Defense Forces/Military Intelligence, *International Financial Aid to the Palestinian Authority Redirected to Terrorist Elements,* June 2002. As of July 18, 2010:
http://www.mfa.gov.il/MFA/MFAArchive/2000_2009/2002/6/International%20Financial%20Aid%20to%20the%20Palestinian%20Aut

Jones, Seth, and Martin Libicki, *How Terrorist Groups End: Lessons for Countering al Qa'ida,* Santa Monica, Calif.: RAND Corporation, MG-741-1-RC, 2008. As of July 6, 2010:
http://www.rand.org/pubs/monographs/MG741-1/

Kagan, Kimberly, "The Anbar Awakening: Displacing al Qaeda from Its Stronghold in Western Iraq," *Iraq Report,* Washington, D.C.: Institute for the Study of War, August 21, 2006–March 30, 2007. As of February 26, 2010:
http://www.understandingwar.org/files/reports/IraqReport03.pdf

Kami, Aseel, and Michael Christie, "Al Qaeda's Iraq Network Replaces Slain Leaders," *Reuters,* May 16, 2010.

Kelly, John F. (Lieutenant General, U.S. Marine Corps), "Foreword," pp. vii–x, in Timothy S. McWilliams (Chief Warrant Officer-4, U.S. Marine Corps Reserve) and Kurtis P. Wheeler (Lieutenant Colonel, U.S. Marine Corps Reserve), eds., *Al Anbar Awakening: Volume I, American Perspectives—U.S. Marines and Counterinsurgency in Iraq, 2004–2009,* Marine Corps University, United States Marine Corps, Quantico, Va.: Marine Corps University Press, 2009.

Kilcullen, David J., "Field Notes on Iraq's Tribal Revolt Against Al-Qa'ida," *CTC Sentinel,* Vol. 1, Issue 11, October 2008.

Kiser, Steve, *Financing Terror: An Analysis and Simulation to Affect Al Qaeda's Financial Infrastructures,* Santa Monica, Calif.: RAND Corporation, RGSD-185, 2005. As of June 29, 2010:
http://www.rand.org/pubs/rgs_dissertations/RGSD185/

Kohlmann, Evan F., "State of the Sunni Insurgency in Iraq: August 2007," New York: The NEFA Foundation, 2007. As of July 18, 2010:
http://nefafoundation.org/miscellaneous/iraqreport0807.pdf

Krueger, Alan B., and Jitka Malečkova, "Education, Poverty and Terrorism: Is There a Causal Connection?" *The Journal of Economic Perspectives,* Vol. 17, No. 4, 2003, pp. 119–144.

Levitt, Steven, and Sudhir Alladi Venkatesh, "An Economic Analysis of a Drug Selling Gang's Finances," *The Quarterly Journal of Economics,* Vol. 115, No. 3, August 2000, pp. 755–789.

Lomborg, Bjorn, "Is Counterterrorism Good Value for Money?" *The Mechanics of Terrorism,* NATO Review, April 2008.

Long, Austin, "The Anbar Awakening," *Survival,* Vol. 50, No. 2, April–May 2008, pp. 67–94.

MacFarland, Sean, and Niel Smith, "Anbar Awakens," *Military Review,* March–April 2008.

Magouirk, Justin, "The Nefarious Helping Hand: Anti-Corruption Campaigns, Social Service Provision, and Terrorism," *Terrorism and Political Violence,* Vol. 20, No. 3, 2008, pp. 356–375.

Makarenko, Tamara, "The Crime–Terror Continuum: Tracing the Interplay Between Transnational Organised Crime and Terrorism," *Global Crime,* Vol. 6, No. 1, February 2004, pp. 129–145.

Malkasian, Carter, "A Thin Blue Line in the Sand," DemocracyJournal.org, Summer 2007.

McWilliams, Timothy S. (Chief Warrant Officer-4, U.S. Marine Corps Reserve), and Kurtis P. Wheeler (Lieutenant Colonel, U.S. Marine Corps Reserve), eds., *Al Anbar Awakening: Volume I, American Perspectives—U.S. Marines and Counterinsurgency in Iraq, 2004–2009*, Marine Corps University, United States Marine Corps, Quantico, Va.: Marine Corps University Press, 2009.

Metz, Steven, and Raymond Millen, "Insurgency in Iraq and Afghanistan: Change and Continuity," prepared for the National Intelligence Council, Carlisle, Pa.: Strategic Studies Institute, U.S. Army War College, 2004.

Miller, Greg, "Bin Laden Hunt Finds Al Qaeda Influx in Pakistan," *Los Angeles Times,* May 20, 2007.

Ministry of Planning and Development Cooperation, *Iraq Living Conditions Survey 2004,* Baghdad, Iraq, 2005. As of January 3, 2010:
http://www.cosit.gov.iq/english/cosit_surveys.php

Multi-National Force–Iraq, Significant Activities (SIGACTS) III Database, Baghdad, Iraq, undated.

Naylor, R. T., "The Insurgent Economy: Black Market Operations of Guerilla Organizations," *Crime, Law and Social Change,* Vol. 20, 1993, pp. 13–51.

Newell, Richard G., and William A. Pizer, "Discounting the Distant Future: How Much Do Uncertain Rates Increase Valuations?" *Journal of Environmental Economics and Management,* Vol. 46, 2003, pp. 52–71.

Office of the Director of National Intelligence, "Letter from al-Zawahiri to al-Zarqawi," ODNI News Release No. 2-05, October 11, 2005. As of March 11, 2010:
http://www.dni.gov/press_releases/20051011_release.htm

Oghanna, Ayman, "Corruption Stemmed at Beiji Refinery—But for How Long?" *Iraq Oil Report* (posted by Alice Fordham), February 16, 2010.

"Oil Smuggler and Al Qaida Supplier Arrested in Bayji," *Al Mashriq Newspaper,* November 22, 2007, as cited in Phil Williams, *Criminals, Militias, and Insurgents: Organized Crime in Iraq,* Carlisle, Pa.: Strategic Studies Institute, U.S. Army War College, June 2009.

Oppel, Richard A., Jr., "Iraq's Insurgency Runs on Stolen Oil Profits," *New York Times,* March 16, 2008.

Passas, Niko, "Terrorism Financing Mechanisms and Policy Dilemmas," in Jeanne K. Giraldo and Harold A. Trinkunas, eds., *Terrorism Financing and State Responses: A Comparative Perspective*, Stanford, Calif.: Stanford University Press, 2007.

Picarelli, John, and Louise Shelly, "Organized Crime and Terrorism," in Jeanne K. Giraldo and Harold A. Trinkunas, eds., *Terrorism Financing and State Responses: A Comparative Perspective*, Stanford, Calif.: Stanford University Press, 2007.

Prober, Joshua, "Accounting for Terror: Debunking the Paradigm of Inexpensive Terrorism," Policy Watch #1041, Washington, D.C.: The Washington Institute for Near East Policy, November 1, 2005.

Roddy, Michael, "Qaeda Confirms Deaths of Leaders in Iraq: Statement," *Reuters*, April 25, 2010.

Russell, Charles, and Bowman Miller, "Profile of a Terrorist," in Lawrence Zelic Freedman and Yonah Alexander, eds., *Perspectives on Terrorism,* Wilmington, Del.: Scholarly Resources Inc., 1983, pp. 45–60 (originally published in *Terrorism: An International Journal,* Vol. 1, No. 1, 1977, pp. 17–34).

Samuels, Lennox, "Al Qaeda Nostra," Newsweek Web Exclusive, May 21, 2008. As of May 7, 2010:
http://www.newsweek.com/id/138085

Security Service (UK MI5), "Al Qaida's Ideology," undated. As of March 22, 2010:
https://www.mi5.gov.uk/output/al-qaidas-ideology.html

Shapiro, Jacob N., "Bureaucratic Terrorists: Al Qa'ida in Iraq's Management and Finances," in Brian Fishman, ed., *Bombers, Bank Accounts and Bleedout: Al-Qa'ida's Road In and Out of Iraq*, U.S. Military Academy, West Point, N.Y.: Combating Terrorism Center, 2008. As of January 2, 2009:
http://www.ctc.usma.edu/harmony/pdf/Sinjar_2_July_23.pdf

———, "Terrorist Organizations' Inefficiencies and Vulnerabilities: A Rational Choice Perspective," in Jeanne K. Giraldo and Harold A. Trinkunas, eds., *Terrorism Financing and State Response: A Comparative Perspective,* Stanford, Calif.: Stanford University Press, 2007a.

———, "The Terrorist's Challenge: Security, Efficiency, Control," Palo Alto, Calif.: Center for International Security and Cooperation, Stanford University, April 26, 2007b. As of July 18, 2010:
http://igcc.ucsd.edu/research/security/DACOR/presentations/Shapiro.pdf

———, "Organizing Terror: Hierarchy and Networks in Covert Organizations," Paper presented at the annual meeting of the American Political Science Association, Marriott Wardman Park, Omni Shoreham, Washington Hilton, Washington, D.C., September 1, 2005a.

Shapiro, Jacob N., and David A. Siegel, "Underfunding in Terrorist Organizations," paper presented at the annual meeting of the American Political Science Association, Marriott Wardman Park, Omni Shoreham, Washington Hilton, Washington, D.C., September 1, 2005b.

Simon, Steven, and Daniel Benjamin, "America and the New Terrorism," *Survival*, Vol. 42, No. 1, Spring 2000, pp. 59–75.

Sly, Liz, "Top Two Al-Qaeda in Iraq Leaders are Dead, Officials Say," *Los Angeles Times,* April 20, 2010.

United Nations World Food Programme, Iraq Country Office, and Central Organization for Statistics and Information Technology, Ministry of Planning and Development Cooperation, Iraq, *Food Security and Vulnerability Analysis in Iraq*, United Nations World Food Programme, 2006.

U.S. Department of Defense, *Measuring Stability and Security in Iraq: September 2009 Report to Congress in Accordance with the Department of Defense Supplemental Appropriations Act 2008 (Section 9204, Public Law 110-252)*, Washington, D.C., November 4, 2009.

———, *Measuring Stability and Security in Iraq: September 2007 Report to Congress in Accordance with the Department of Defense Appropriations Act 2007 (Section 9010, Public Law 109-289),* Washington, D.C., September 14, 2007.

Valentim, Joice, and Jose Mauricio Prado, "Social Discount Rates," Working Paper, University of São Paulo, Brazil, and IMT Lucca Institute for Advanced Studies, Italy, May 6, 2008.

Warner, John T., and Saul Pleeter, "The Personal Discount Rate: Evidence from Military Downsizing Programs," *The American Economic Review,* Vol. 91, No. 1, March 2001, pp. 33–53.

Weaver, Mary Anne, "The Short, Violent Life of Abu Musab al-Zarqawi," *The Atlantic,* July/August 2006. As of January 19, 2010: http://www.theatlantic.com/doc/200607/zarqawi/3

Whitlock, Craig, "Al-Qaeda Masters Terrorism on the Cheap," *The Washington Post,* August 24, 2008.

Williams, Phil, *Criminals, Militias, and Insurgents: Organized Crime In Iraq,* Carlisle, Pa.: Strategic Studies Institute, U.S. Army War College, June 2009.

Williamson, Oliver, *Markets and Hierarchies: Analysis and Antitrust Implications,* New York: Free Press, 1975.